HANDBOOK OF SOLAR AND WIND ENERGY

HANDBOOK OF SOLAR AND WIND ENERGY

A Cahners Special Report

Floyd Hickok

CAHNERS BOOKS INTERNATIONAL, INC.
221 Columbus Avenue
Boston, Massachusetts 02116

Library of Congress Cataloging in Publication Data

Hickok, Floyd, 1907-
Handbook of solar and wind energy.

(A Cahners special report)
1. Solar energy--Handbooks, manuals, etc.
2. Wind power--Handbooks, manuals, etc. I. Title.
TJ810.H5 621.4'5 75-22495
ISBN 0-8436-0159-0

ISBN: 0-8436-0159-0

Printed in the United States of America

CONTENTS

PREFACE

Glowing promises appear in the press every day about the abundance of solar energy. It has been well established that the sun and the wind could supply all of an industrial society's energy needs. If to these resources were added a few other energy inputs such as geothermal, tides, and hydro, there would be no such thing as an energy shortage or an energy crisis. Sooner than most people think, it would be possible to depend on these "non-fossil" resources without burning another drop of oil or speck of coal and without adding another one of the controversial fission reactors.

Many people ask, When and how will the promises of solar and other non-fossil energy be fulfilled? Others think the promises are empty and efforts to pursue them will be a tragic waste. However, it is the contention here that non-fossil energy will become practical and will find its rightful place as fast as entrepreneurs find it to be a resource to be exploited.

The most likely entrepreneur to exploit non-fossil energy is the one who is already in the energy business. The utility company is already in the energy business, and it is also one that is being especially squeezed by the oil crisis. Therefore, attention is given to the relation between solar and wind energy and the utility business. Practical applications cannot be described because none exist at present. Suggested, however, are lines of thought for utility managers, architects, and engineers to pursue if the text convinces them that non-fossil energy is a money-making resource.

This report offers an in-depth survey of the state of the art in solar and other non-fossil energy and a timetable of probable development. Professionals of any industry will find this survey a comprehensive introduction to the various energy resources.

HANDBOOK OF SOLAR AND WIND ENERGY

1

NON-FOSSIL ENERGY—A PRACTICAL IDEA

NON-FOSSIL ENERGY RESOURCES ARE RIGHT AT HAND

Some of the non-fossil resources, such as geothermal energy, are indigenous to a locality. Others, such as the sun and the wind, are everywhere. In either case the resources are right at hand to be exploited—it requires only the invention of the proper technology.

These non-fossil resources are theoretically attractive because:

—they are inexhaustible
—they are non-polluting
—they are free, except for the cost of conversion
—the cost of conversion is stable and easy to predict. The cost of energy from these resources has no inflationary effect
—the amount of energy available from these resources is enough to make all peoples self-sufficient
—full exploitation of these resources can lead to the complete elimination of fossil burning. These valuable raw materials can then be used for the manufacture of goods.

The non-fossil energy resources are there for the taking. It requires only the will to act. When Erich Farber, noted University of Florida solar energy pioneer, was asked how soon the sun could supply a big portion of the energy budget, he replied, "Just as soon as we want it."

HOW MUCH ENERGY WILL BE NEEDED?

This Report projects energy requirements out to the year 2020. This date will probably see the retirement of all fossil burning systems now in operation or in the planning stage. By 1980 electric utilities will be in position to judge whether to phase out fossil burners and move into alternative prime resources. The forty years from 1980 to 2020 is about the maximum anyone can propose for a planning basis.

One way to predict the future energy needs is to base the estimate on trends in per capita consumption. In 1975 the U.S. per capita consumption of energy was in the order of 400 million BTU per year and growing at an annual rate of 5%. How long this yearly increase will go on, given the freedom to expand, is difficult to estimate. A Ford Foundation report (A Time to Choose, Energy Policy Report) points out that there are saturation points beyond which the amount of consumption grows very little and consumption will eventually stabilize. It is assumed in this Report that by 2020 the per capita consumption will be fairly stable at 600 million BTU per year.

The effect of increased per capita consumption will be compounded by the growth of population. It is impossible to offer a reliable estimate. Robert Drake (V-P, Combustion Engineering, Inc., Stamford, Ct. 06902) points out: "The long-range planner must face the unpleasant fact that the difference between the low estimate and the high estimate (of population growth) in only 30 years time amounts to almost one-half the present population of the United States." That amount of population spread compounded with the spread of per capita estimates builds an enormous uncertainty in the estimates of energy needs.

It is necessary, however, to take some reasonable number and "act as if" that is the way things are going to be. The number used here is a population of 300 million by 2020.

Using these guesses the national energy demand in 2020 will be 180 quadrillion BTU.

Another way to predict the energy needs of the future is to base the estimate on categories. The broad groups usually defined for this type of estimating are residential, commercial, industrial and transportation. Table 1-1 is a listing of the probable demands of these groups limited to electrical consumption only.

If, as some estimators assume, electric generation will consume 50% of the energy budget by 2020, and if the total consumption at that time is 180 quadrillion BTU, then the electric utility consumption will be 90 quadrillion BTU distributed among residential, commercial and industrial at about the ratio of Table 1-1.

Table 1-1. National Energy Consumption by Categories, Electricity Only. Quadrillion BTU.

	1975	1985	2000	2020
Residential	7.5	13 to 13.5	10.8 to 22.5	13 to 44
Commercial	8.8	6 to 6.9	18.4 to 45.2	30 to 180
Industrial	9	9.3 to 18.3	15.6 to 45.6	30 to 100
Transportation*				

The utility manager will find these gross national estimates useful if he compares them with the statistics of his own franchise area. If his numbers compare rather closely with the national 1975 numbers, he may extrapolate from the national estimates. Note that the residential estimates are based on 68 million occupied dwellings in 1975. The sizes of commercial and industrial units vary too much from area to area to draw a reliable comparison.

CAN NON-FOSSIL ENERGY SUPPLY THE NEEDS?

What energy resources will supply the enormous energy demands of the coming years? Conventional thinking says it will come from domestic oil, imported oil, shale oil, coal, hydropower and nuclear fission. Figure 1-1 is a diagram showing the expected energy consumption and distribution of fuels. Variations in this diagram, attributed to Gaucher, have appeared in a great many of the papers and books on energy. The small area occupied by non-fossil energy reflects the conventional thinking about energy resources.

---

*Transportation is one of the unpredictables for the electric utility. To the extent that mass transit generates its own electricity it is a factor in the gross national energy budget, but that factor does not affect the utility's plans. If electric cars, powered by either flywheel or battery, come into common use, there would be some impact on utility plans. Fritz Kalhammer of Electric Power Research Institute has estimated that electric cars will have an increase impact of 1.8% in electric generation in 2000.

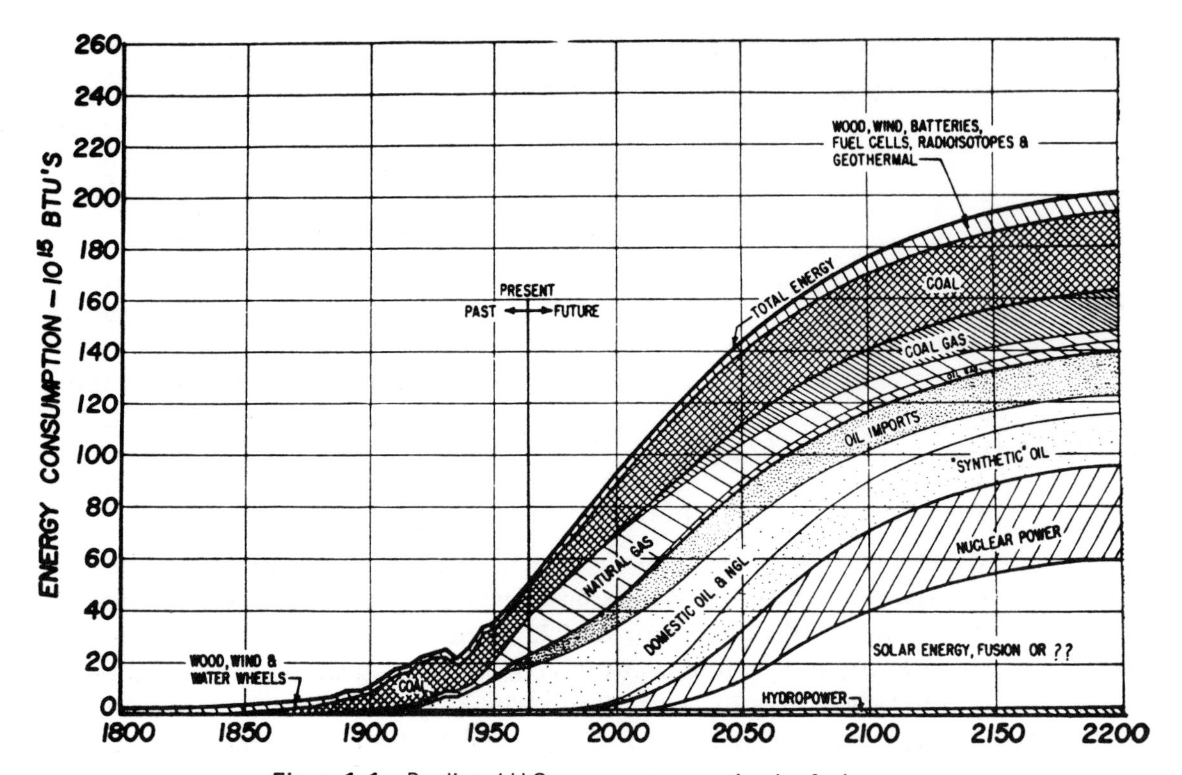

Figure 1-1. Predicted U.S. energy consumption by fuel types.

Should these non-fossil resources be taken seriously? William Cherry of NASA says that the total U.S. electrical needs for 1990 could be met by devoting 1% of the land area to solar cell arrays. The Meinels of the University of Arizona say that 13,000 square miles of southwest desert could generate all of the electricity needed by this country and northern Mexico for the next hundred years. Alfred Eggers, Jr., of the National Science Foundation, says that wind power could provide, yearly, 300 billion Kwhr off the shores of New England, 190 billion off the Texas gulf coast, and 400 billion off the Aleutian chain. Huge amounts of power are recoverable from the winds of Texas and the great plains. Considering that the total electrical consumption in the nation for 1974 was 2000 billion Kwhr, the wind energy resource is impressive.

Advocates of non-fossil energy tend to think of it as a producer of electricity. At present electricity does only 10% of the work of the nation; it will never do all of the work. If non-fossil resources could supply all of the needed energy, how could they supply these BTU's other than by electricity? The answer: by direct conversion to heat or by the production of hydrogen through the electrolysis of water or thermochemical conversion of water.

Hydrogen as an energy resource is not discussed technically in this Report because it is considered a secondary resource. The aim here is to examine the technologies of the primary non-fossil resources, except nuclear, to see what the utilities can do about them.

Think Small versus the Energy Factory

Utility people traditionally think big and bigger. This attitude is ingrained both in the nature of their business and in natural law. It is necessary to build larger generating plants to meet increasing demands, and the Carnot law requires bigger plants to get better efficiency. Hence, when looking at non-fossil resources, the tendency is to look for ways to generate hundreds or thousands of megawatts. Anything less than a megawatt is not worth considering.

Thinking big is also engrained in the nation's political philosophies. For example, Public Law 93438, which sets up the new Nuclear Regulatory Commission as well as the new Energy Research and Development Administration, says that the Regulatory Commission is authorized to adopt policies that will encourage the location of nuclear power reactors and related fuel cycle facilities on nuclear center sites.

The nuclear center site is an example of what G. G. Leeth of General Electric Company, Santa Barbara, Calif., calls the "energy factory." A factory system is characterized by intense concentration of effort towards the production of the largest possible amount of product at the hardest pinch-penny policy. The factory system works wonders for the production of a large amount of inexpensive hardware, and it probably will do the same for energy production.

It is suggested here that to think effectively about non-fossil energy requires the think-small approach. The energy factory system encourages the kind of thinking that would try to put all the production into one plant, whereas the non-fossil systems lead to the search for a little bit here and a little bit there until the whole is found.

It is suggested here that everyone, especially the utility people, evaluate the energy factory system before it becomes the sole policy. Can it truly supply all the energy needed, is it flexible enough, is it not dangerous to depend on total energy from a few localities? The objection is that the energy factory system could eliminate non-fossil systems before they have a chance to develop.

THREE TYPES OF NON-FOSSIL ENERGY

Non-fossil energy can be converted from three resources, namely, those that depend on the sun, those that depend on the nature of matter, and those that depend on the earth itself.

The conversion technologies may be summarized as follows:

- Solar Energy
  - Direct Conversion
    - Low Quality Heat
    - High Quality Heat
    - Solar Voltaic
  - Indirect Conversion
    - Wind
    - Ocean Thermal Differences
    - Biological Conversion
- Properties of Matter
  - Fusion
  - Fuel Cells
  - Batteries
  - Miscellaneous
- Earth Energy
  - Geothermal
  - Tides

2

SOLAR ENERGY

HOW MUCH SOLAR ENERGY IS THERE?

Each advocate of solar energy seeks to emphasize the amount reaching the earth by describing it in some dramatic fashion. For example, the amount falling on Lake Erie is enough to supply the needs of the nation.

The total amount reaching the earth is enormous, but the amount available for conversion to useable form is small. Most of the land area is occupied by living and commercial space, or by agriculture and forest space, and a certain amount of space, of a quantity as yet unknown, must be left nearly untouched to maintain the ecological balance. Even so, an area of about 40,000 square miles would provide all of the energy needs predicted for 2020. This figure is derived by assuming solar radiation at 1 Kw per square meter and computing the conversion efficiency at 20% with 2500 hours of sunlight. Some 500,000 square miles of the nation are devoted to farming. Solar energy, thought of as a ratio to other land use, does not seem such a bad idea.

Some scientists have been studying the meaning of this amazing quantity of energy for half a century or more. When some of them dramatize their findings, they may be letting their frustrations show because they know the energy is there, yet so few are willing to reach out and take it.

In personal terms, solar energy interests everyone. For example, 900 square feet of roof in the Southwest consisting of solar devices working at 10% efficiency would generate 80 Kwhr during a clear spring day. The average home uses about 25 Kwhr per day, leaving 55 Kwhr available to charge an electric car. Some estimators say that 30 Kw will drive a car for 200 miles. With such rosy promises, no wonder solar engineers are impatient.

When the sun is directly overhead at noon, the solar flux on a horizontal surface energizes at the rate of 1 Kwhr, or 3413 BTU, per hour per square meter.

That value is the maximum possible rate. Conversion efficiencies take a percentage off this theoretical maximum. For instance, some places on the earth never have the sun directly overhead, and everywhere on earth deviations from the maximum vary widely with the seasons and time of day and cloud cover and haze.

The sun rarely if ever strikes a horizontal plate at normal incidence. The flux rate is almost always less than maximum. For example, at 40 degrees north latitude the flux rate at 9:00 A.M. on January 21 is 896 BTU on a horizontal plate, or 26% of maximum, and at noon on the same day it is 1771 BTU or 52% of maximum. (Flux rate per square meter.)

The flux rate can be improved by tilting the plate to approximately the latitude of the locality. In this case the same hours show, respectively, 1845 BTU or 54% of maximum, and 3143 BTU or 92% of maximum.

The flux rate can be further improved by rotating the plate so that it always faces normal to the sun. In this case the same hours show, respectively, 2581 BTU or 76% of maximum, and 3175 BTU or 93% of maximum. Rotating the plate is expensive, and solar engineers avoid it except in special cases.

The flux rate varies constantly from sunrise to sunset. The total energy falling on a surface is the value that interests the engineer. At 40 degrees north latitude the daily total energy falling on a plate tilted to 40 degrees is 24,000 BTU per square meter on June 21 and 17,647 BTU on December 21.

These numbers imply a clear, cloudless day.

HOW TO MEASURE SOLAR ENERGY

The Weather Bureau uses three solar radiation instrument types: the sunshine duration device, the pyroheliometer and the pyranometer.

The sunshine duration device commonly used is a Campbell-Stokes sunshine recorder. This instrument is a spherical lens about three inches in diameter. It focuses direct radiation onto a strip of paper. The sun's heat produced at the focal point discolors the paper. As the sun sweeps through the azimuth the focal point travels and traces a line on the paper. The length of the trace measures the amount of sunshine in the day.

Photovoltaic instruments are also used for sunshine duration measurement. There are 162 stations taking photovoltaic measurements, but they are not under Weather Bureau cognizance.

The pyroheliometer measures the normal incidence or direct radiation. Sunlight is admitted to the front end of the barrel, at the far end of which is a thermopile. Diffuse radiation is prevented from influencing the thermopile by the barrel. An alt-azimuth mount automatically tracks the sun. Instantaneous readings, in langleys, are taken at certain times during the days that are not totally cloudy.

As of July, 1974, the Weather Bureau was taking pyroheliometer measurements at only six stations.

The pyranometer measures the total direct and diffuse radiation. The sensor is mounted on a plane parallel with the earth. Looking up at the whole globe of the sky, it measures the "global" radiation. The measurements are continuous, and the data are usually entered on a strip chart recorder.

The source of Weather Bureau archived solar information is National Climatic Center, Federal Building, Asheville, N.C. 28801.

Most of the commercially available solar radiation instruments used in this country are manufactured by Eppley Laboratory (Newport, R.I. 02840). This company makes pyroheliometers, pyranometers and infrared radiometers. It also makes instruments for calibration of solar devices with artificial light sources. The sensitive element of the Eppley pyroheliometer is a thermopile, and of their pyranometer it is a thermocouple.

Spectrolab also manufactures radiation instruments.

With the expanding interest in solar energy, there is continuing effort to increase the long-term stability and reduce the cost of solar instruments. Norris and Tricket report (Solar Energy, Vol. 12, pp 231-253) the development of a pyranometer having a sensitive element consisting of a thermopile in the shape of a rosette with the cold junction at the perimeter where heat sinking is easily effective.

The sensitive element of most instruments for measuring solar radiation either generates or modifies an electric current. In the Eppley pyranometer the temperature difference between an absorbing black plate and a reflecting white plate causes a change in current through the thermocouple. The current change is proportional to the heat absorbed by the black plate. The absorbed heat is expressed in calories. The plates of the instrument have a measured area. The output of the instrument is therefore expressed as a heat rate in calories per square centimeter.

The international unit of solar energy is the langley, which is 1 calorie per square centimeter. Instruments for meteorological use are calibrated in langleys. The primary standards for the International Pyroheliometer Scale are maintained at the National Physical Laboratory, London. In this country a standard, calibrated against the primary standard, is maintained by Eppley. One langley per minute is approximately 697.3 watts per square meter per minute, or 0.3172 BTU per square foot per hour.

Solar engineers use the term "insolation" to express the idea of solar flux rate. The definition found in the National Bureau of Standards' proposed standard for solar collectors is, "Insolation is the rate of solar radiation received by a surface." Solar engineers usually prefer to express the rate in BTU per square foot rather than in langleys.

As is well known, radiation from the sun covers a broad spectrum from X-rays through ultra violet, visible light and the invisible infrared. The atmosphere filters out most of the radiation except that in the visible spectrum. The wave lengths in the visible spectrum range from 0.3 to 3 microns. The sun's rays upon entering the atmosphere are dispersed or reflected by particles of dust and droplets of water; the total skylight has a diffuse component of about 10%. When the direct component and the diffuse component are measured together, as they are in the pyranometer, the sum represents the total insolation.

Meteorologists refer to the visible spectrum as short-wave radiation. This classification contrasts with the radiation band from 4 to 50 microns, which is called long-wave radiation. The long-wave band is also the infrared or heat band.

As most of the infrared band from the sun is filtered out before it reaches the earth's surface, the long-wave component of energy at the surface is of terrestrial origin. There are two sources of this component: one, the downward emission from the gases of the atmosphere, especially water vapor and carbon dioxide, and two, the upward emission by natural surfaces and gases. There are thus 4 components of radiation: direct short-wave, diffuse short-wave, incoming long-wave and outgoing long-wave. In the emerging demand for more accurate determination of the solar insolation, Eppley and others are concentrating on developing the means of measuring each of these components.

When solar measurements are collected and analyzed over a period of time, they become an archive of solar radiation data. Unfortunately, most of the collected data is global radiation collected from a horizontal sensor. As most solar devices are slanted, it would be desirable to have data on a slanted surface. There are conversion factors available for slanted surfaces, but direct measurements would be more desirable for the practical designer. A further bleak outlook is the fact that the Weather Bureau has not published solar radiation data since September, 1972, because the long term stability of many instruments has broken down with a consequent loss of sensitivity. Considering the need for accurate and varied data for second and third generation solar devices, help from the government in the measurements area is not very promising. The Weather Bureau "hopes" that money will be available to set up forty new stations using new instrumentation.

Air mass data is necessary for the refined application of solar energy. It is disappointing to see no effort to make this measurement along with insolation measurements. Air mass is a physicist's scale ranging from 0 through 5. It is 0 in space, and at the earth's surface on a very clear day it is 1.

Lacking hard data for a locality solar engineers rely on archive data for general application, as does this Report.

One good reference book for general radiation data is "Report No. 21. World Distribution of Solar Radiation" by Lof, Duffie and Smith. It is available from the College of Engineering, University of Wisconsin. This book reports the average global radiation in langleys for each of 12 months from most of the weather stations of the world. The book includes 12 world maps in color with isotherms showing the monthly distribution of solar energy.

The ASHRAE Guide (American Society of Heating, Refrigerating and Air-Conditioning Engineers, Inc., 345 East 47th St., New York, N.Y. 10017) gives tables for different latitudes showing insolation values for various angles of measurement. ASHRAE, one of the engineering societies most active in solar energy promotion, publishes a number of very valuable books for solar engineers.

The National Bureau of Standards (Washington, D.C. 20234) publishes six insolation tables based on latitude in their "Proposed Standard Method of Testing for Rating Solar Collectors." These tables, based on ASHRAE data, are a helpful source of general information.

3

FLAT PLATE COLLECTORS

FLAT PLATE COLLECTORS ARE HEAT ABSORBERS

As light is an electromagnetic wave, it would be nice to be able to convert solar radiation directly into electricity. Raytheon Company has done extensive research on a device called the "rectenna," which is capable of converting very short radio waves directly into electric power. W. C. Brown (Journal of Microwave Power, December, 1970) shows that efficiencies greater than 80% should be expected from a fully developed rectenna system. If a rectenna could be developed for each of the frequencies over the visible spectrum, a remarkable solar conversion device would be available. No such device is on the horizon, however.

The practical converter now available is the black surface absorber. The absorber works by taking up the short-wave visible radiation and converting it to long-wave infrared energy. The absorber dissipates the infrared heat energy either by re-radiation or by convection into an absorbing metal. The flat plate collector seeks to prevent re-radiation and enhance absorption. The black surface absorbs about 95% of the light striking it at the normal angle.

Flat plate collectors, being the simplest of all devices for converting solar energy, have been made by motivated handymen. They consist of a flat, black absorbing surface, usually painted metal, attached to pipes through which a fluid, usually water, circulates to drain away the heat. To prevent the heat from being dissipated by wind or re-radiation, the absorber and pipes are placed in an insulated frame covered by glass or plastic. The temperature of the fluid in the collector pipes rises to 100 degrees or more above ambient temperature. Since the sun shines only a part of the time when heat is needed, heat storage is necessary to make the system practical. A tank of water is usually used for storage, although some handcrafted systems have used a bin of rocks.

Some Flat Plate Collector Manufacturers

The simplicity of the flat plate collector has attracted a number of entrepreneurs seeking to break into the solar energy business.

One such manufacturer is Revere Copper and Brass Incorporated (P.O. Box 151, Rome, N.Y. 13440). Revere's collector is an extension of their copper clad laminated panel, which was developed for the construction industry. Each panel measures 2 × 8 feet and carries from 2 to 5 tubes. After determining the amount of heat required by the building, the designer is able to specify the number of panels and number of tubes per panel by consulting the company's technical literature. When all manufacturers supply good design information in tabular and graphic form, as Revere has done with its technical literature, solar energy for buildings will become more acceptable because architectural consulting engineers count on using manufacturing design data. Figure 3-1, a schematic diagram for a solar heating system, is an example of graphic design information.

The National Science Foundation has let a number of contracts to install proof of concept solar heating systems in several parts of the country.

General Electric Co. (P.O. Box 8661, Philadelphia, Pa. 19101) has one of the contracts. They have installed 150 panels, size 4 × 8 feet, on the roof of the Grover Cleveland Junior High School in Dorchester, Mass. These panels consist of a black heat-absorbing surface beneath two sheets of clear "Lexon" plastic. A tubing network inside the black surface is filled with a water/anti-freeze solution. The heated solution is pumped through the tubing network to a pair of special solar heat exchangers that work in conjunction with two of the school's 10 conventional heating units. This experimental installation is expected to supply about 20% of the school's heating needs.

The solar heating system for the Timonium Elementary School in Baltimore County, Maryland, was installed by AAI Corporation (P.O. Box 6767, Baltimore, Md. 21204). AAI, sensing the potential in solar heat after their experience with the school, has undertaken a development program to improve their position. They have announced a "Roof-Top Concentrator," and tests during development suggest the probability of producing solar heat for under $2 per million BTU.

International Environment Corporation (129 Halstead Ave., Mamaroneck, N.Y.) offers flat plate collectors with a proprietary black coating having a high

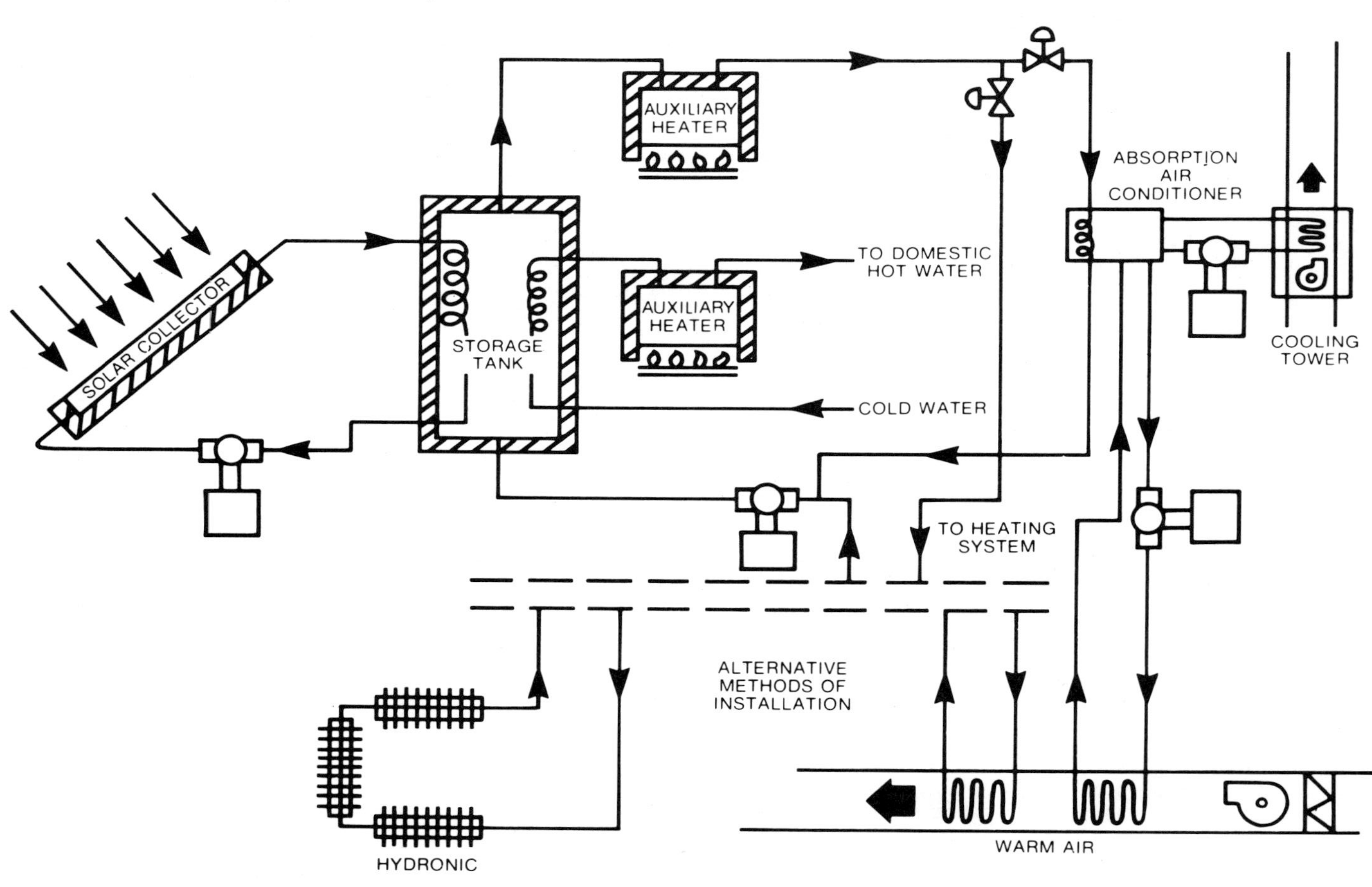

Figure 3-1. Solar energy schematic diagram.

absorbance to emission ratio. The glazing cover is two sheets of tempered glass with a U value of 0.53. This company's solar panel work is an extension of its experience with radiant heating and cooling systems.

Kalwall Corporation (1111 Candia Rd., Manchester, N.H. 03103) is a manufacturer of translucent wall panels and skylights through which light and solar heat enter buildings. The panels are of fiberglass. Kalwall has extended its fiberglass capabilities to the production of special fiberglass glazing for flat plate collectors. This company has a series of solar systems undergoing tests in anticipation of entering the solar energy market.

Up until the government stepped in with demonstration projects the only systems built were hand crafted or sponsored by university research departments. So far, anyone interested in installing a system has been forced to handle the entire design, procurement and installation. With the high initial cost, low confidence in performance and inability to hand craft a system, there has been very little incentive to adopt solar energy. The idea of saving several hundred dollars a year in heating costs is appealing, but for most people there has been no practical way to enjoy the benefits.

With the entrance of commercial enterprise into the market that picture is likely to change.

Standards Are Important

The National Bureau of Standards has recently released a proposed standard for testing the rating of solar collectors. This specification describes in detail how to test a solar panel for efficiency. It will enable independent test organizations to rate commercial flat plate collectors—a prerequisite for successfully marketing these devices. It will help manufacturers to evaluate their own designs. It will bring standardization to a field that is presently helter-skelter. This specification contains a list of definitions that will become standard in the solar energy trade.

The push by the government and the engineering societies to establish standards is important, but the customer may still have problems with the meaning of efficiency. Collectors tested by the NBS specification may show 70 to 80% efficiency. The heat obtained by the building may be substantially less. Here is why: Efficiency is a ratio of output to input. If the output is the energy rate at the exit port of the collector and the input is the radiation incidence falling on the collector, the exit energy may be quite close to the applied energy. However, the collector works into a storage tank, and the tank works into a building, which is in itself a rather good storage system. It is easily possible for the storage system to be unable to accept much, if any, additional input. The storage water may be at or close to the temperature of the collector fluid. In this case the collector keeps on receiving input energy, but the system efficiency is low.

Most solar engineers settle on 50% system efficiency as a good average. In winter, and in summer with air conditioning, the efficiency is better than 50% because almost all of the heat collected is used. In spring and fall the efficiency is less because more heat is collected than is used.

System efficiency standards would be very difficult to devise and apply unless a way could be found to store and use all the energy collected.

HOW GOOD IS THE COST STORY?

Ball-park estimates for adding a flat plate collector system to a home, new or old, are $3000 to $5000. The system should pay for itself in 10 to 15 years with the owner getting at least 10 years "free" heat from it.

G. O. G. Lof of Colorado State University and R. A. Tybout of Ohio State University have done extensive analyses of system costs. In their work (reported in Solar Energy, Vol. 16, pp 9-18) they find that systems combining space heating, space cooling and water heating are competitive with conventional energy in most parts of the country.

Lof and Tybout found that they could ascribe costs in dollars per Kwhr or BTU for any of the variables, such as collector size, collector tilt angle, number of covers, or storage volume for heating, for cooling, and for heating and cooling combined. In every case the annual cost for combined service was less than for either alone.

Of particular importance was the discovery that the curves drawn from the data differed substantially from locality to locality. Figure 3-2 shows curves for Albuquerque and Miami in terms of collector size. Note that in Miami the heating cost increases almost linearly with collector size because the need for heating is minimal.

Lof and Tybout have computed the costs for 8 U.S. cities. Some of the results are shown in Table 3-1. The costs in this table are based on 20 year

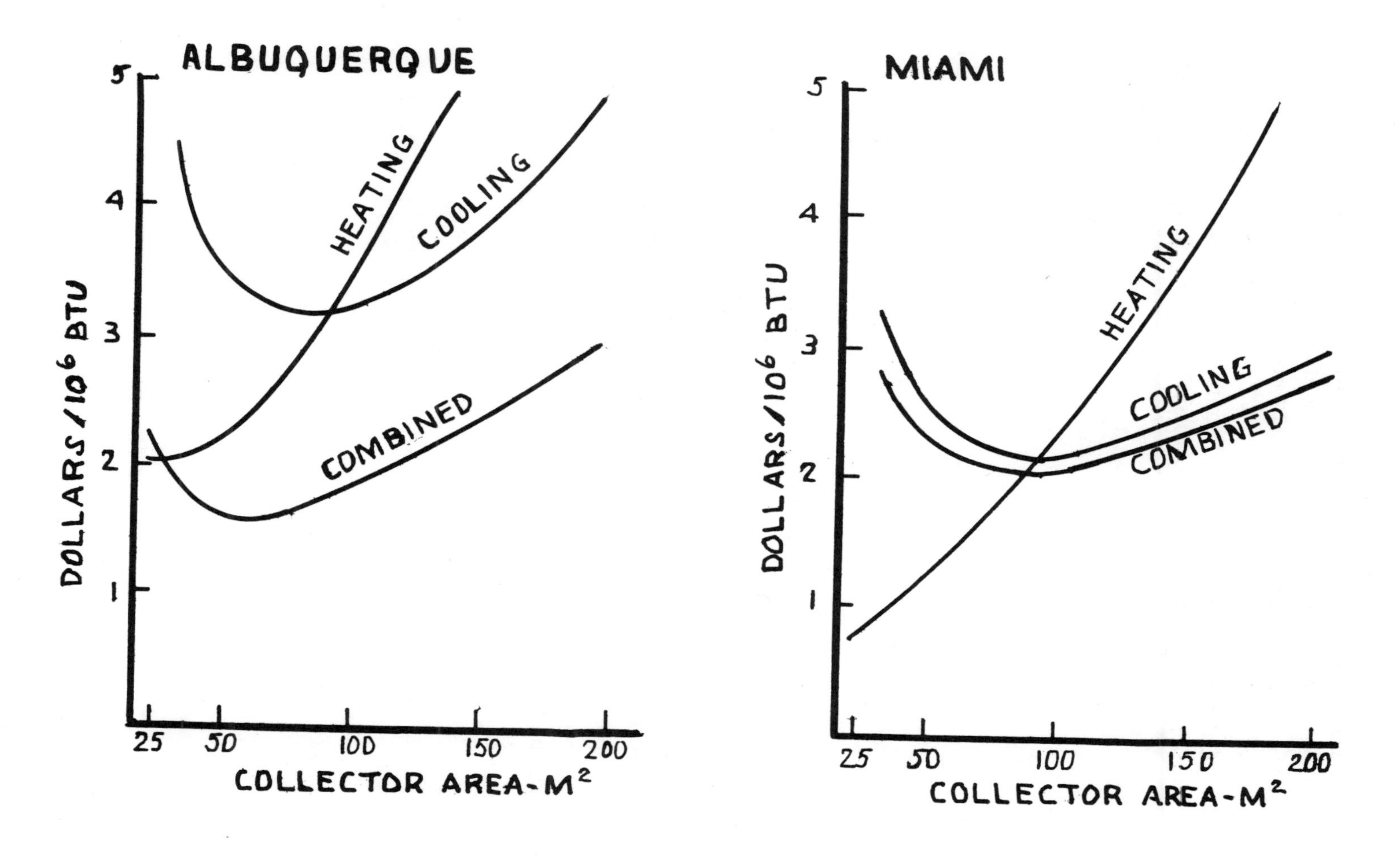

Figure 3–2. Collector size versus energy output, Albuquerque and Miami.

Table 3-1. Eight-City Flat Plate System Costs
(Combined space heating and cooling
and hot water. In 1972 dollars.)

	Collector Area M^2	% Load by Solar	Cost $/$10^6$ BTU	Cost $/100 Kwhr	Electricity $/100 Kwhr
Albuquerque	48.4	63	1.73	0.59	1.58
Miami	96.7	60	2.2	0.73	1.66
Charleston	96.7	68	2.47	0.84	1.44
Phoenix	96.7	33	1.71	0.58	1.73
Omaha	96.7	59	2.48	0.85	1.11
Boston	96.7	65	3.07	1.05	1.79
Santa Maria	24.2	52	2.45	0.84	1.48
Seattle	48.4	43	3.79	1.29	0.78

amortization at 8% and on collector cost at $21.53 per square meter ($2 per square foot). These computations are aimed at optimizing a solar system in conjunction with conventional energy.

Lof has recommended that the solar system should be designed to supply not more than 80% of the building needs with conventional heat supplying the rest. Note from the table that in most cases for optimum cost advantage the heat load is usually about 60% of total. Going to 80% pushes the advantages a little bit but does not go too far.

4

SOLAR ENERGY FOR SELF-SUFFICIENCY

A MODEL COMMUNITY'S FLAT PLATE COLLECTOR RESOURCES

In 1972 solar energy advocates were saying that flat plate collectors could not supply all the heating needs of a building, at least on an economical basis compared with conventional fuel. Suppose the issue were phrased on a different track. Suppose the question were, can a community supply all of its energy needs from flat plate collectors?

Consider a model community consisting of 10,000 dwelling units, 100 commercial buildings, one light industry, one shopping center and 4 schools and public buildings. It is located on the 40th parallel and receives 2500 hours of sunlight a year.

Each of these dwelling units could support a roof collector of 1000 square feet. In this community such a collector slanted to the optimum angle is exposed to an insolation value of 480 million BTU per year. Most units could find space on the ground for an additional 1000 square feet of collector, or an additional 480 million BTU per year.

It is almost always assumed that flat plate collectors should be slanted to about the degree of latitude. Vertical surfaces of buildings offer a neglected opportunity, however. At latitude 40 degrees the insolation value of a vertical plate has some interesting characteristics. The vertical plate collects more energy in winter than in summer, whereas the slanted plate collects more in summer than in winter. They complement each other. Used together one makes up for deficiencies in the other.

On most houses the south wall could provide at least 25% of the roof collector area, or 250 square feet. The vertical surfaces could contribute an additional 72 million BTU per year. The total insolation value for this typical dwelling unit is 1.032 billion BTU per year. At 50% efficiency this dwelling could collect

516 million BTU per year. The 10,000 dwellings could therefore collect 5.16 trillion BTU per year.

The commercial buildings could each accommodate 3000 square feet of slanted plates for a total energy collection of 720 million BTU at 50% efficiency, or 72 billion for all the commercial buildings.

The light industry occupies a single story building of 10,000 square feet. It could accommodate 8000 square feet of slanted collector. In addition, there is a three-acre parking lot, which could accommodate at least 120,000 square feet of slanted collector. This 128 thousand square feet of collector could supply 30.72 billion BTU per year at 50% efficiency.

The shopping center consists of 300 thousand square feet of buildings and 15 acres of parking lot. This area could accommodate at least 800 thousand square feet of collector. It could produce 184 billion BTU at 50% efficiency.

The public buildings could accommodate at least 35,000 square feet of collector (more if parking lots were used), or 8.4 billion BTU at 50% efficiency.

In summary:

	10^9 BTU/yr
Residences	5160
Commercial	72
Industry	30.72
Shopping Center	184
Public Buildings	8.4
Total	5455.12

Assuming that the residences use 200 million BTU per year each for space conditioning and water heating, the 10,000 units require 2 trillion BTU per year. For estimating purposes, assume that all the other buildings use 120 thousand BTU per year per square foot for the same service. Their total needs are 102 billion BTU. The needs for the entire community for heating, cooling and hot water are 2.102 trillion BTU per year.

If the flat plate collector capacity is 5.45512 trillion BTU and the community's demand for heat is 2.102 trillion BTU, the community has a surplus of 3.35312 trillion BTU.

The community has a population of 30,000. At the 1975 per capita consumption of 390 million BTU per year (which may be high for this community since it has no heavy industry) the community's energy demand is 11.7 trillion BTU per year. This quantity is total demand. It includes the heat budget of 2.102 trillion, the electricity budget, the transportation budget, etc. The flat plate collectors of the community are therefore short of supplying total needs by 6.25 trillion BTU.

It must be remembered that at least half of the consumed energy in most communities is represented by stored energy in the form of purchased goods. Since any article of commerce has required energy to produce it, when it is purchased it becomes imported energy. Therefore, only half of the 11.7 BTU represents local consumption. In this light, the flat plate collectors of this model community are capable of supplying approximately their total energy needs.

Assuming a per capita consumption of 600 million BTU per year in 2020 and the same population, the community demand would be 18 trillion BTU. Assuming the same collector output and an energy import of 50% in the form of goods, the community would be short 3.55 trillion BTU in 2020. The deficit could be made up from a 340-acre solar farm.

This community covers 12,000 acres. The solar farm would require 2.8% of the land. However, this community is probably 95% developed, and the undeveloped part probably consists of parks and woodlots and wetlands necessary to preserve the ecological balance and quality of life. It may be difficult to find 340 acres suitable for a solar farm.

It is obvious that this scenario for community self-sufficiency is too crude to be acceptable by modern standards. But consider these factors: The pressure for exploitation of non-fossil energy may become so articulate that ideas of acceptability may change rapidly. The old problem of cost may change many people's minds. There are systems on the horizon whose efficiency is far greater than 50%. While the scenario is improbable, the point is, a community does not have to import its energy from a thousand miles away to have self-sufficiency.

Sandia's Solar Community

While the scenario for community self-sufficiency may seem like science fiction, the basic idea is taken seriously at Sandia Laboratories (Albuquerque, N.M. 87115), where $2 million in NSF and AEC funds were spent in fiscal '74 and '75 on a "Solar Community" project.

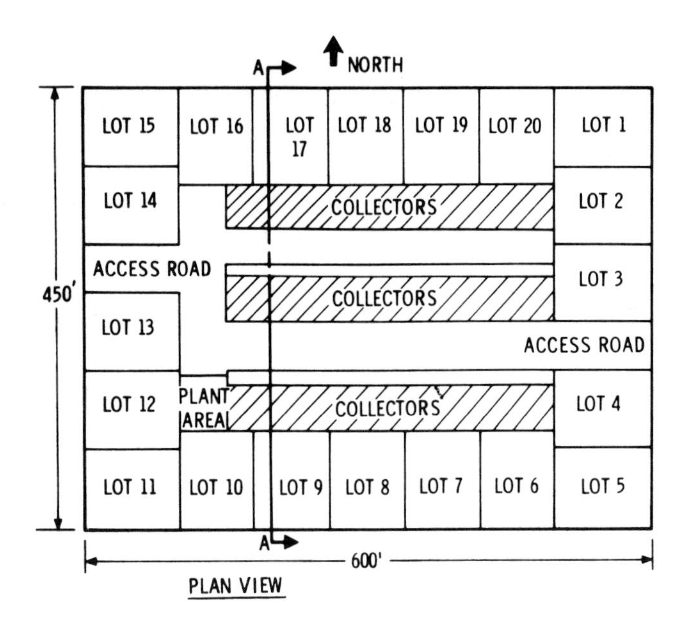

Figure 4-1. Sandia Laboratories' model solar community.

This project involves the collection and storage of solar energy and distribution via water pipes and power lines to meet the needs of a community for electricity, heating and air conditioning and hot water. This project differs from the model community suggested here by the use of central collection and storage instead of a system dispersed among the buildings.

Residences and commercial buildings require both high quality energy in the form of electricity and low quality energy in the form of heat. It would be nice to disperse the waste heat from a generator to supply the low quality heat needs of buildings. This cascading principle is in fact used in large cities that have a central business district steam generator. The spent steam is delivered to local office buildings for heat and air conditioning. Only local buildings can benefit because it is too expensive to deliver low quality heat beyond a moderate distance.

Sandia engineers have adopted the cascading idea and have combined it with the concept of a unit community. They escape the penalty of long distance delivery of low quality heat by clustering the community around a central solar system. A typical solar community is shown in Figure 4-1.

The collector selected by Sandia is the half-moon reflector described here in the subsection on high quality heat. They have chosen the parabolic cylinder

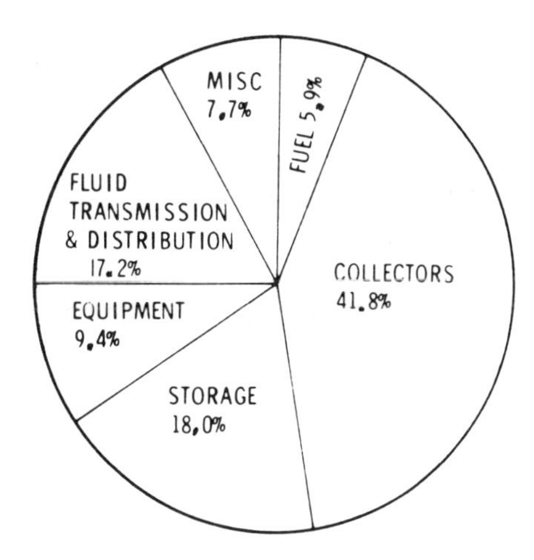

Figure 4-2. Cost distribution, Sandia Laboratories' model solar community.

configuration rather than the full half-cylinder cross section. The total collector area is 20,000 square feet.

Storage is by water in a pressurized tank for the heat from the collector and by a separate tank of water for the low quality heat to be delivered to the buildings. Water in the pressurized tank rises to 365 to 450 degrees F.

Electricity is generated in a Rankine cycle turbine using an organic fluid such as n-pentane or toluene. The overall efficiency is 11%.

Heating and cooling may be pumped from the central station to the various buildings either by forced air or water.

Sandia's studies have shown that there is more advantage in the collection and storage of energy on the basis of a small subdivision than for a single house or for a much larger community. The maximum size of the subdivision has not been established. The size they have chosen to study is a 20-dwelling community. The projected fuel savings are 60 to 70%.

Sandia has attempted to derive a cost estimate of energy for the solar community, but has run into the same difficulty encountered by everyone else. There is no reliable prediction of collector costs; it might run from $2 to $10 a square foot. It is impossible to predict government policy, money costs and conventional fuel prices. Using the best guesses possible, they come to the conclusion that energy from the solar community would cost about twice that of conventional energy. Figure 4-2 shows their division of costs.

Sandia's project has met with widespread enthusiasm, encouragement, invitations to present the idea to various groups, and inquiries as to how the idea might have applications to other activities. The local utility has expressed interest in helping to build a demonstration solar community in the late '70's.

5

AIR CONDITIONING, THE KEY TO DOMESTIC SOLAR ENERGY

SPACE COOLING, CAN THE FLAT PLATE COLLECTOR DO IT?

Solar heating systems are much more ready for the market than cooling systems. Significant problems remain in the understanding of mechanisms, in the design and operation of cost-effective hardware, and in the determination of economic performance. There are no cooling devices presently manufactured that are specifically designed to match the characteristics of the flat plate collector system.

In fact, there are some who wonder whether the flat plate collector can ever operate properly with a cooling device.

Since it is widely accepted that if solar energy is to be cost-effective it must be combined with heating, cooling and water heating, many companies are pushing solar air conditioning development. Several different devices are being vigorously defended. It will take a few years yet to determine which, if any, will work out, or what niche, if any, each will occupy.

A survey of the leading contenders follows.

The Absorption System, a Refrigeration Cycle

Conventional air conditioners work on the principle that an evaporating gas acts as a refrigerant. They re-compress the gas by means of an electrically driven pump. There are, however, refrigerators and air conditioners that use heat to evaporate the fluid and condensers to convert the gas back to a fluid. This refrigerator is called the absorption type.

The logical thought is to stick to the refrigerator idea but go to the absorption type instead of the compressor because solar heat could be used as the evaporating energy.

To follow this path requires the selection of refrigerating fluids that will evaporate at the temperatures attainable from flat plate collectors. The two contenders are aqua ammonia and lithium bromide.

An aqua ammonia air conditioner has been built and is in operation at the University of Florida's solar demonstration house.

Lithium bromide air conditioners are commercially available (from Arkla). They operate from a gas flame, but some models can be modified to operate from flat plate collector heat.

There are a couple of drawbacks to absorption solar coolers:

—The condenser cycle requires a large cooling source, usually a tank of water. A complete solar house would then require two large tanks, one for heating, one for cooling.
—The acceptability of flat plate collector low quality heat is marginal. Absorption systems can begin to refrigerate at inputs of 180 degrees F, which is high for the flat plate collector except on the hottest days. The absorption system coefficient of performance is poor at this temperature.

Desiccant Systems

The desiccant principle is similar to the absorption principle in that the evaporation of a liquid produces cooling. In this case the liquid is water. If water evaporates into air, the air is cooled. Evaporation of $1\frac{1}{2}$ gallons of water is equivalent to 1 ton of air conditioning.

A space cooling cycle is established by drawing the warm, moist air from the building over a desiccant bed and then blowing the dried air into the building through a humidifier to cool it. Air into the building can be cooled to 60 degrees F by this method. When the desiccant becomes saturated, solar heat is used to regenerate it. Two dessicant beds are necessary, one for the absorption phase and one for the drying phase.

The desiccant system is well suited to the single family dwelling, but it is not so good for large buildings because the weight of the desiccant goes up linearly with system size. However, the system looks good as a dehumidifier for large buildings if used with a regular air conditioner.

Such an application will be tried out on the upper 25 floors of the Citicorp Center being built in New York City. The south-facing tower at the top of the building will support 20,000 square feet of flat plate collectors. Since a part of the function of the compressor air conditioner is to dehumidify the air, the solar desiccant system will relieve the main unit of that load. It is expected that the system will save about 5% of the cost of air conditioning in this section of the building.

New York's Consolidated Edison Company participated financially and technically with the preliminary feasibility studies of this system.

Putting a Waterbed on the Roof

The "Skytherm system of cooling and heating consists of a pond of water on the roof. The pond container has sliding covers. In winter the panels are moved to expose the water during the day to let the sun warm the water; the water is covered at night to prevent escape of the heat. In summer the water is covered during the day and exposed at night to allow the heat to escape through radiation and evaporation. The inside of the house is cooled or heated as required by re-radiation from the roof pool. The weight of the water on the roof produces less building stress than does a bin of rocks or a tank of water.

The concept of using a "waterbed" heat sink as one surface of a building is best suited to the Southwest. A test house has been successfully built at Atascadero, California.

Rankine Cycle Engines for Cooling

The Rankine cycle engine can be used to convert solar energy into mechanical energy, which in turn drives a compressor type air conditioner. Several companies have been working on this engine, including The Garrett Corporation, Hittman Associates, United Aircraft, TECO and Barber-Nichols Corporation.

In the next chapter solar systems capable of developing many hundreds of degrees will be described. The Rankine cycle engine can be designed to accommodate a wide range of temperatures from those of the flat plate collector to those of much higher degree.

When not driving the air conditioner the engine can operate an electric generator to feed power to the building.

The fact that this engine can work on low quality heat means that, in many industrial processes where much heat is wasted because there is no way to retrieve it, substantial amounts of power can be recovered. This spin-off from the solar air conditioning effort could become quite significant. The drive for the conservation of energy gives much attention to insulation and avoiding waste. An effort to retrieve waste heat could be equally important.

Rankine cycle solar air conditioners should be ready to market at competitive prices in perhaps 3 years according to TECO engineers.

The above survey shows that the cooling system winner has not yet emerged. There are many offerings. Perhaps each is equally good. Most of them have not gone beyond the prototype stage. If the flat plate collector cannot really move ahead unless it combines both heating and cooling, and if solar cooling is still in the development stage, then flat plate collectors will have to wait awhile.

Perhaps the situation can best be summed up in the words of Sean Wellesley-Miller of M. I. T.: "The temptation is to go for 'brute force' solutions that will be capable of serving all building types under all climatic conditions. Yet, if we do find such a generalized solution, which is still in doubt, it will almost certainly be as excessively expensive as its mechanical predecessors were excessively energy consuming. Personally, I believe that if solar cooling is to be genuinely successful we will have to develop and use a variety of technical approaches depending on local climate and building type." (From a paper presented at the Workshop for Solar Cooling for Buildings, February, 1974.)

6

WHAT'S NEXT?

THE NEXT GENERATION SOLAR COLLECTORS

Second generation solar collectors may be ready to market before first generation models get beyond the demonstration phase. New developments are being pushed rapidly.

Concentrators Raise Collector Temperatures

The heat generated by a square foot of collector can be increased if more than a square foot of solar flux is concentrated upon it. The method of concentrating by means of trapezoidal grooves is now under intense development.

The theory of the concentrator is explained by K. G. Hollands in Solar Energy, Vol. 13, pp. 149-163, 1971. The basis of the idea is that a trapezoidal groove with sides sloping at 30 degrees reflects all incoming rays to the bottom of the groove. The geometry is shown in Figure 6-1. If the ratio of the bottom width to side height is optimally selected, the last ray hitting the top edge of the side will be reflected to just the opposite edge of the base. The sides of the trapezoid must be highly reflective.

The University of Houston, Department of Mechanical Engineering (3801 Collen Blvd., Houston, Texas 77004), has been studying this type of collector under contract with the National Science Foundation. A design they have developed is shown in Figure 6-2. This design has a base to side ratio of 1.75. With a selective coating on the absorber plate at the base of the groove, the predicted efficiency is between 40 and 55% at a plate temperature of 275 degrees F for all seasonal noon times. The concentrator collector is 10 to 20% less efficient than the flat plate collector, but it develops a higher temperature.

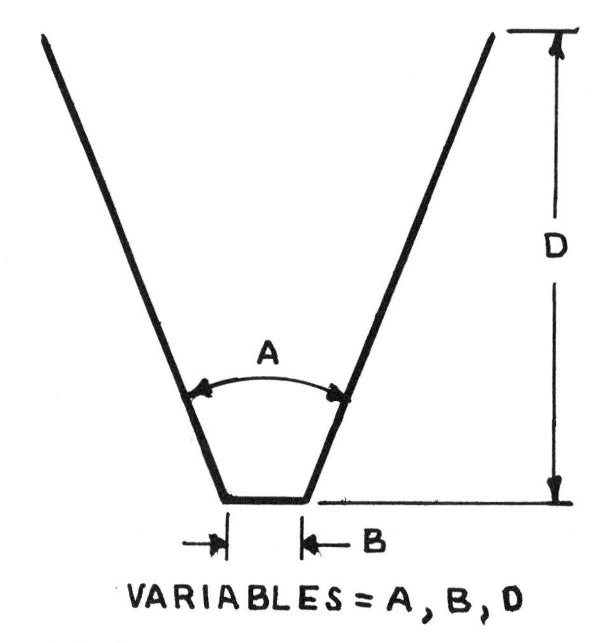

Figure 6-1. Cross-sectional geometry of concentrator collector.

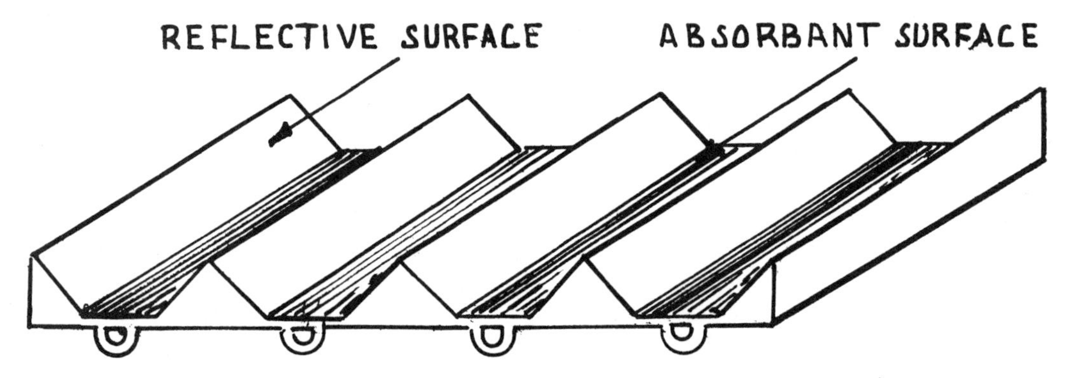

Figure 6-2. University of Houston grooved concentrator collector.

The concentrator collector shares with the flat plate collector the advantage of being able to harvest diffuse as well as direct light. Both are able to collect some energy even on cloudy days.

The concentrator collector will be essential for both the absorption and the Rankine cycle air conditioners.

The concentrator collector is partially directive. It should be mounted so that the grooves run east and west, and the array should be pivoted to about the latitude of the installation. The array should be mounted on a pivot so that the slant can be tilted several times a year. The optimum efficiency is attained when the noon sun is normal to the base. The array should be kept at a slant such that the base is always ±20 degrees to the normal.

Superior Surfaces Are Coming

An ideal flat plate collector cover would admit all of the visible light and reject or contain all of the infrared radiation. An ideal absorber surface would absorb all of the visible light and re-radiate all of it as infrared. Either or both of these surfaces would improve the efficiency of the collector. Research is progressing on these materials. Since cost per square foot is the marginal factor for the flat plate collector system, the materials will have a cost struggle to justify their added efficiency.

FLAT PLATE COLLECTORS, AN ARCHITECTURAL PROBLEM

Even if flat plate collectors become technically and economically attractive, they still have to be put on a building (usually). In the end they become an architectural problem. Listed below are some of the issues the architects must resolve to make solar energy acceptable:

—The First Cost Syndrome. The market place has forced architects to design for least first cost, leaving the owner and tenants to achieve interior comfort by wasteful, brute force methods. Since the original owner has no intention, usually, of keeping the building for the amortized life of the solar equipment, he is first cost motivated. Can the architect invert this situation?

—Insulation Skimping. Everyone agrees that the first place to begin to conserve energy is with insulation. But insulation increases the first cost. There is a beneficial trade-off. Since insulation reduces energy demands, better insulation reduces the size and cost of the solar system.

—Reconciliation with Present Practices. Building practices are rigidly maintained. Building codes and trade union customs are ingrained. The architectural design of solar systems must be reconciled with current building practices, with present hardware, with present legal, financial and lending customs.

—Technological Architecture. The architect is accustomed to the practice of making a general design and then turning over to a consultant the task of making the building stand up and be properly lighted, heated and cooled. Some people think that solar energy will require architects to become more technically oriented. Technological architecture means designing a building so as to harmonize the climate with the best technical selection of materials and processes for using solar energy.

—Living with Nature. The protesting, communal, back-to-nature movements of the '60's spawned a school of architects and engineers with a live-with-nature ideal. Architecturally this idea means designing the building and energy system together in terms of the local environment. The movement to design with nature seems so much like emphasizing the obvious it may win out over the present brute force control-of-nature philosophy.

A study has shown that optimized climatic building design could reduce the season heating load in New York by 50% and the cooling load by 70%. This philosophy of design could reduce the size of the solar plant to the point where its first cost would no longer be a barrier.

STORAGE IS MORE DIFFICULT THAN IT SEEMS

The methods of storing solar heat are easy to find. Each method has its own technical problems, which usually can be resolved by technical solutions.

What cannot be overcome are the limits imposed by thermodynamic laws.

These limits are considered here in relation to water storage, although the principles apply to any type of heat storage.

If it takes 100 BTU to raise a certain volume of water by 1 degree, the same heat would raise twice the volume by only one-half degree. Over the course of the day the solar collectors can generate only so much heat. If the water volume is too large, its heat gain may be only a few degrees, or not enough to supply heat to the building.

There is an optimum relation between size of storage and size of collector for both economic and thermodynamic reasons. If storage is too small the total output of the collector will not be absorbed. If it is too large the water acts as a diffuser, not as storage.

Lof and Tybout have found that the heat storage volume varies with locality. In Phoenix it is 48.8 kg/m^2, whereas in Boston it is 73.2 kg/m^2.

The other thermodynamic problem has to do with rate of heat transfer. In the low quality heat system of the flat plate collector the problem is one of equilibrium temperature. For satisfactory heat transfer the collector output needs to be at least 15 degrees F above the storage temperature. If the cooling system must operate from temperatures of around 200 degrees F, it may be difficult to attain storage temperatures high enough to operate the system.

Thermodynamic limits of storage as much as anything determine that the practical system should supply only about 80% of real needs. If all of the collected energy could be inexpensively stored for a long period and eventually all of it used, it would be practical to install a system with 100% capacity. As this ideal is unattainable with any of the systems so far described, some lesser capacity must be compromised, and the power company must supply the balance.

A SUMMARY OF FLAT PLATE COLLECTOR STATUS

The National Science Foundation's fiscal '75 budget for research on heating and cooling of buildings with flat plate collectors was $17 million. The aim is to make flat plate collector heating available commercially before 1978 and cooling by 1980.

Flat plate collector heating is already technically practical, but it has high first cost. This high first cost might come down with large scale production, which can only be achieved if there is an expanding market.

The market is presently beset by inertia. National Science Foundation is sponsoring studies looking into marketing problems, building code problems and other restraints on commercial acceptance.

It is likely that by the end of the decade some large community will be built as a solar demonstration project.

It is quite possible that low quality heat is the limiting factor restricting acceptance of flat plate collectors. This characteristic limits storage and provides marginal performance for cooling. The quality of heat from the flat plate collector is just not quite good enough to do all it is supposed to do. If the system does not quite live up to its promises, that fault could kill its market potential.

It may be that the flat plate collector will not be fully acceptable until it is combined with the concentrator. If so, the marketing date will have to slip a bit. The practical concentrator type flat plate collector may not be ready before 1980, about the same time as the cooling system.

7

DIRECT CONVERSION TO HIGH QUALITY HEAT

HOW TO INTENSIFY SOLAR ENERGY

By contriving to bend the sun's rays from a wide area so that they occupy a small area the heat in the small area can be intensified. Some contrivances raise the temperature to several thousand degrees. These devices are called solar furnaces and are used for specialized research. Other contrivances raise the temperature to 500 degrees F up to 1500 degrees F. These devices are the ones of particular interest to solar engineers because that range of temperature is common to conventional electric generators.

The devices of interest here are either reflectors or focusers. They have a common problem. They are efficient only when their attitude is normal to the sun. This characteristic requires that they track the sun. Automatic tracking equipment immediately raises the cost. These devices do not respond to diffuse light.

Reflectors

The parabolic reflector is one possible configuration. The parabola has the characteristic that rays hitting any point on the surface are reflected to the focal point. A conversion device at the focal point would be able to utilize the sun's energy.

Another idea would place an array of perhaps hundreds of flat mirrors on sloping ground and adjust the mirrors so that they all reflected onto a mirror at the top of a tower.

The half-moon groove

The cylindrical reflector with an energy-absorbing pipe along the axis in either the half-cylinder or the parabolic contour seems to be the most promising of the reflector configurations. It resembles the concentrator to some extent, except that the absorbing surface is at the front instead of at the back of the groove.

The parabolic reflector is illustrated in cross section in Figure 7-1. Sunlight is reflected from the surface onto a pipe carrying an absorbing fluid at the parabolic axis. The pipe is surrounded by a glass tube. The back side of the tube is a reflector, and the side toward the parabola has a window to admit the rays to the central pipe. Energy is recovered by drawing the fluid out to some sort of heat engine.

Investigators working with the grooved concentrators have chosen to orient the array so that the grooves run in the east-west direction. No tracking is required with this mounting, except an occasional adjustment of the slant. At first it seems that the half-moon reflector ought to be mounted in the same way. However, Sandia Laboratories, during their work on "The Solar Community," found

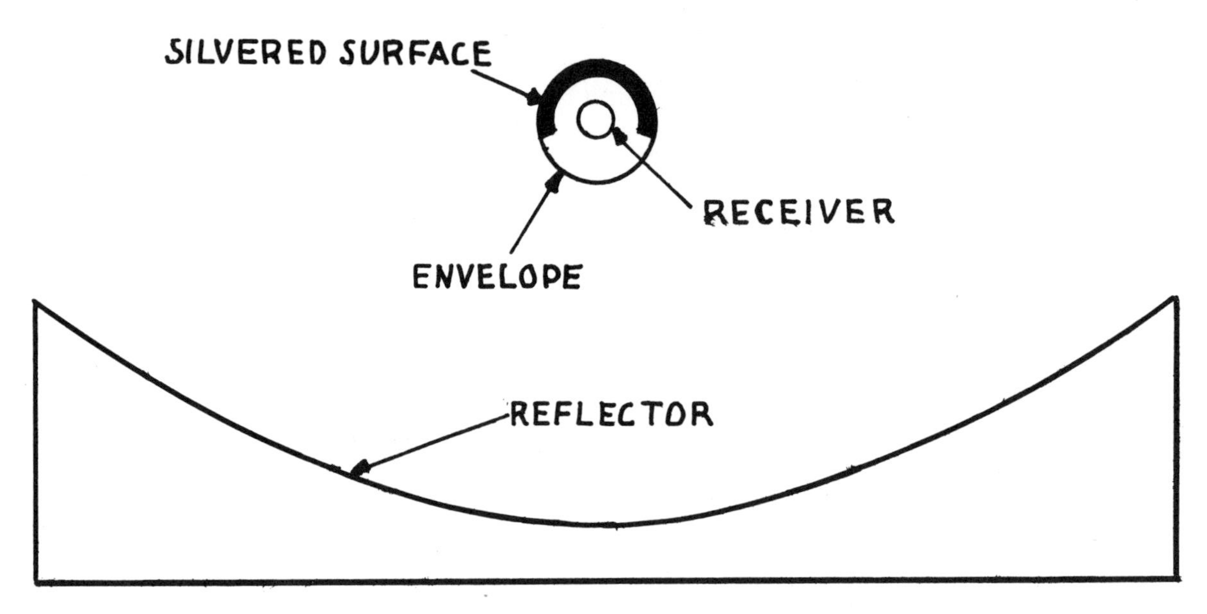

Figure 7-1. Cross-section of parabolic reflector collector.

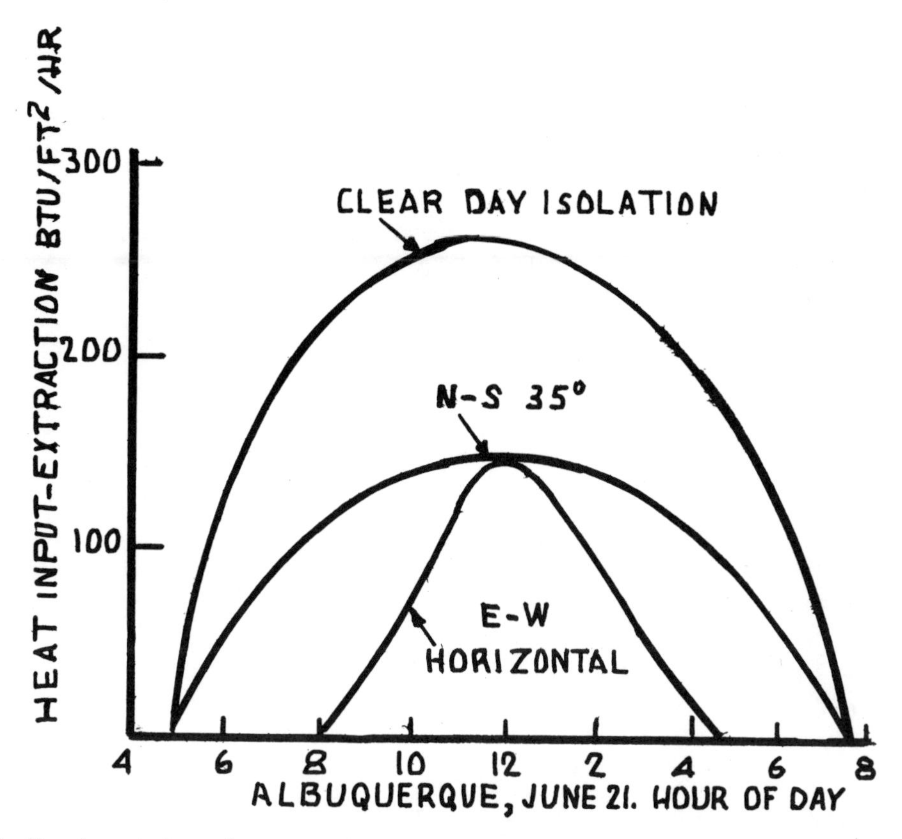

Figure 7-2. Hourly variations of energy collected by east-west pointing and north-south pointing reflector collectors.

that the preferred mounting was with the axis in a slanted north-south direction. The slant should be about the latitude of the installation. This mounting requires automatic tracking in the east-west direction, but the additional energy collected in the north-south orientation makes the tracking expense worth it. Figure 7-2 shows the difference in the amount of energy collected in a day between the tracking north-south mounting and the east-west mounting. In mounting the array the collectors have to be spaced far enough apart so that one does not shade another at any time during the tracking.

Sandia has found, however, that they can violate the non-shading rule without too much penalty. With spacing the width of the collector there is an abrupt reduction in energy collection in early morning and late evening due to adjacent collector shading, but the simplicity of the tracking mechanism and the reduction of land area, piping, pumping and insulation make this sacrifice worth it.

For the maximum absorption of energy the glass tube would be evacuated. Sandia has found, however, that for their purposes the expense of the evacuating equipment and its operation does not pay for the gain. At least their preliminary thinking is along this line. The point needs to be substantiated by further research. Other unresolved questions in this area include: What is the effect of varying the size of the glass tube and the absorbing pipe? Should a reflecting coating really be placed on the back side of the tube? Is the tube really necessary at all? What is the best flow rate of the absorbing fluid?

Focusers

The lens is an obvious contrivance for focusing a wide area of light onto a small spot. The problem is cost. A glass lens large enough to be of any value would have an impossible price tag. Collectors at $12 a square foot already dominate the cost of a solar system.

The cost problem can be overcome by the use of Fresnel lenses. These lenses are pressed from plastic and can be made in large sizes for a few dollars a square foot.

A lens system would require automatic tracking.

THE MEINELS SUGGEST A SOLAR FARM

Aden and Marjorie Meinel of the University of Arizona have suggested a system for attaining high quality heat from the half-moon reflector. They would connect a large number of these collectors in series-parallel and use the output to drive a conventional steam generator.

Aden Meinel favors the viewpoint that solar energy is more likely to come into use through evolution than through revolution. He sees an acceptable system as one most like the ones in present use. The collector system is well within current technology, and the power generator is one already brought to a high state of development.

The innovative idea advanced by the Meinels is to coat the interior pipe with a selective optical thin film. This coating is highly absorptive in the visible and highly reflective in the infrared spectrum. The infrared is reflected inward so that the transfer fluid (liquid sodium) in the pipe accumulates heat to as much as a thousand degrees F.

Combining the losses from the solar concentrator and the steam generator, the efficiency of the system is about 30%. The Meinels expect it to produce electricity for about 5.3 mills per Kwhr at the bus bar in 1972 dollars.

The capacity of the Meinel system is 226,000 Kw per square mile. A 1000 Mw plant would require 4.5 square miles of land, or 5.24 million square meters of collectors. The Meinels were aiming at a collector cost of $60 per square meter to attain the 5.3 mills per Kwhr. A 1000 megawatt collector system would cost $314 million.

Dr. Meinel, who lives in the desert Southwest and sees thousands of square miles of sun-baked earth lying fallow, would like to see a few of those thousands put to the conversion of energy. He likes to think of his idea as a solar farm, which harvests energy instead of wheat. The yield per acre (at 5.3 mills) would be about $9500, which any wheat farmer would be glad to take.

The solar farm is an energy factory. It would be worthwhile to consider a small scale application. Since his collectors are linear in cost and efficiency, their size is not important.

The possibility of using this high quality heat directly should also be considered. It is well known that the fewer the conversion transfers the better.

What Would the Model Community Do with a Reflector Collector?

The reflector collector is too bulky and awkward to be mounted on a house roof. It cannot take the place of the flat plate collector for that service. If the community has a few acres that can be spared, it could install a solar farm for the harvesting of electricity.

The possibility of using the Meinel system instead of flat plate collectors in a solar community application should be studied. It might be suitable for an apartment complex.

The high quality heat of the reflector collector would be suited to a multistoried building, as it could be installed on the roof of this type of building.

Unfortunately, neither the economics of a small scale Meinel system nor its adaptability to a variety of climates has been studied, so that a valid judgement about the general value of the system is not possible.

FLAT REFLECTORS BECOME A FOCUSING SYSTEM

If a flat mirror reflects the sun upon a nearby object, the object is bathed in light. However, if the mirror reflects the sun upon a screen a long distance away, an image of the sun appears on the screen. That is, at an appropriate distance a mirror is a focusing device. The intensity of the focused image is a function of the size of the mirror.

This property of mirrors is the basis of a solar energy system being investigated by the University of Houston with RANN support. (Alvin F. Hildebrandt, Solar Energy Research Laboratory, University of Houston, Houston, Texas 77004.)

This system of converting solar energy is usually called the tower concept. A boiler is mounted at the top of a tower, and a large number of mirrors on the ground reflect the sun's image onto the boiler, generating steam, which is piped to a steam generator at the bottom of the tower.

The mirrors are attached to alt-azimuth mounts. They automatically track the sun by means of a sensor feedback system. There is a computer override on the feedback system to position the mirrors in the morning and to continue tracking during periods of cloudiness. The mirrors must be spaced so that none shadows another or blocks the reflection of another.

The flat mirror approach has been selected over the focusing mirror partly for economic reasons. It is much cheaper to mass produce the flat mirror. It is expected that the heliostats, or combined mirror and mounts, can be mass produced for about $30 per square meter. The present mirrors are back-silvered glass.

A typical 10-megawatt plant would have a tower 1600 feet high and mirrors dispersed around it over an area of a square mile. Economics of scale determine the size of the plant. If the system is too small, the ratio of the cost of the heliostats to the rest of the system becomes too great. The minimum economical size is a system with a tower height of about 300 feet. These plants are suitable only in regions of bright sun with little haze, such as the desert southwest.

The solar tower is an energy factory system.

Temperature easily attainable is in the range of 700 to 1000 degrees F. The power density in the boiler ranges from 0.4 to 4 megawatts per square meter. The system is capable of concentration ratios of 1000 to 2000.

With the temperature attainable in the solar tower it is possible to split methane and water (the "eve-adam" process) to produce a gas consisting of H_2 and CO. The gas would be piped to the customer, who burns it, extracts the heat, converts the gas back to methane and water and pipes the methane back to the solar converter. This process is an alternative to electrolysis of water, and it permits solar energy instead of fusion energy to be used to produce a combustible gas.

Cost estimates for a solar tower plant are in the order of $750 per Kw. A proof of concept plant is expected to be completed by 1979.

HIGH QUALITY HEAT AND ENERGY SELF-SUFFICIENCY

No studies are available to this Report on the storage of high temperature heat to match solar energy systems. Since low quality heat and its matching storage of the flat plate collector system do not economically supply total needs, the question arises whether the same rule applies to high quality heat from the reflector collector.

The low quality heat system is weak because it cannot fill up a sufficiently large storage and because it has difficulty pumping the higher level of heat into storage.

Note the situation that prevails with the reflector collector. When the solar energy first begins to warm up the collector in the morning, the transfer fluid quickly rises to its minimum temperature, say 700 degrees F. By noon it has reached its maximum, say 1000 degrees F.

Suppose the storage fluid were some, as yet unknown, material whose properties were to freeze at, say, 300 degrees F and to boil at 700 degrees F.

If the storage volume were sufficiently large it would take a long time, say two weeks, for the temperature to rise to, say, 650 degrees. Since the minimum input temperature is 700 degrees, there is always energy being pumped into storage. If there were an exceptionally long period of high input and low usage, the storage temperature might reach the boiling point, but even then energy would continue to be pumped in most of the day because the maximum temperature is much higher than the boiling point. This scheme assumes a high quality insulation such that the heat loss would be no more than two or three per cent a month.

It is assumed that the building heating and cooling systems are similar to those used with the flat plate collector. This system draws from the high temperature storage, which is never below 300 degrees and is almost always above 500 degrees. The building conditioning system seems to be looking into a heat sink of infinite capacity. The energy source is self-sufficient.

Just as with the flat plate collector, there would be some optimum combination of collector and storage size for best suitability to the climate and for best economy.

Whether there are any materials that would allow such a system to work is not known. The idea does suggest a direction in which some thinking ought to go. The thermodynamic limits of the high quality heat reflector collector may have some surprises.

The flat plate collector system is good for supplying, economically, up to 80% of the demand. Could the reflector collector act as a peaking system for the remaining 20%?

8

ELECTRICITY DIRECTLY FROM THE SUN

SOLAR VOLTAIC CELLS ARE ON THE HORIZON

Solar cells generate electricity when light strikes them. Solar cells do exist, and they are in practical use. They were invented for the space program, but they are now found in many terrestrial applications. They are customarily used in places or situations where conventional power is unavailable. These space-type solar cells work. Why, then, are they not used to generate electricity for home and industry?

The problem is money. The present space-type solar cells cost from $30 to $200 a watt. Therefore, nearly everyone wrestling with the energy crisis dismisses the solar cell with a shrug. Sure it is a nice idea, but at $30,000 a kilowatt? Solar cells at that price will never make a significant contribution to the energy needs.

The investigations leading to this Report suggest that these gloomy prospects are wrong. Solar installations costing less than nuclear plants are on the horizon with a better time scale than breeder fission, coal gasification and some other alternatives being offered with vociferous argument.

SECTION 1. THE SILICON SOLAR CELL

Silicon Cells Are Just Plain Sand

The raw material for this cell is silicon dioxide, or just plain sand. Silicon is the second most plentiful element of the earth. Silicon is completely innocuous. If a silicon array were on the roof and the house burned down, the silicon would simply turn back into sand.

The sand to begin with costs about 0.6¢ a kilogram. Why does this stuff, which is so plentiful and cheap in nature, cost so much in the form of a solar array? The answer is in the manufacturing process.

How To Make a Silicon Solar Cell

The silicon solar cell is an outgrowth of the semi-conductor industry, and for this application the silicon must be very pure. Sand is melted in an electric arc and subjected to hydrogen reduction at high temperatures. This batch of poly-crystalline silicon is the purest material made by any industry. Having passed through this scrupulous purifying process, it now sells for $78 a kilogram.

The next step is to grow a single artificial crystal of pure silicon. The process has been known for 150 years. It consists of dipping a seed crystal of the material into the surface of a pot of the element in the molten state and slowly and painstakingly withdrawing the crystal as it grows. This process produces a boule or single crystal of pure silicon about 3 inches in diameter and 2 feet long.

Flat sheets are needed for the solar cells. They are obtained by taking cross-sections from the boule. The ends of the boule are cut off, and the rod of silicon is cut into slices by an abrasive wheel process. The slices are about 0.012" thick; if they were any thinner they would break in the processing. For best absorption of solar energy the wafer ought to be about 0.004" thick. A thicker slice puts a higher requirement on the silicon.

When the butcher slices a pound of salami he gets a pound of slices. When a technician slices a pound of silicon boule he gets only about 8% of it in slices. To get 5 pounds of slices about 60 pounds of material has to be wasted. The materials cost alone is now over $2000 a Kw.

The polycrystalline silicon could probably be made cheaper. For the solar cell it does not have to be as pure as for the semi-conductor. Perhaps the crystals could be grown more efficiently, but there is a tremendous investment in equipment, which cannot be discarded lightly. Perhaps slicing techniques could be improved, but there is not much chance of making the slices thinner than 0.012" thick.

The cost of the solar cell crystal needs to come down to about $50 a Kw to be practical in today's electrical world. The prospect of getting from $2000 down to $50 by simple improvement is not very glamorous.

Space Age Silicon Cells at Work

Each wafer produces about one-half volt with a current density equivalent to an efficiency of 13 to 16% of the solar insolation. The cell output increases with increased light intensity and decreases with increased temperature. At 100 mW/cm^2 ($1\ Kw/M^2$) solar intensity and a cell temperature of 77 degrees F the short-circuit current is 25 to 30 ma and the open circuit voltage is 0.55 to 0.6 volt. At maximum power output the power decreases at about 0.5% for each degree C temperature rise. If the cell is operated at a voltage below the maximum power point, the output will be insensitive to temperature until the temperature rises sufficiently to bring the maximum power point down to the operating voltage.

The cells are normally assembled into modules and wired in series to produce a standard output. The wafers are mounted on a printed circuit board and encapsulated to make a module. The modules are then wired in whatever series-parallel arrangement will provide the required voltage and current for the installation.

The module is typically wired for 12-volt output because it is normally used to charge a 12-volt lead-acid storage battery. At a normal operating voltage a typical array of modules delivers 0.2 ampere at the early morning sunlight intensity of 20 mW/cm^2 and 1.1 amperes at a noonday sun intensity of 100 mW/cm^2.

These solar arrays are used with marker buoys and navigational aids, sailboats, telephone repeaters, portable communications, automated weather monitoring, forestry lookout towers, railroad crossing warnings, electrical highway signs, and so on.

Note that silicon cells share with flat plate collectors the ability to convert solar energy even on cloudy days. They respond whenever there is light. Note in Figure 8-1 that the power output in heavy haze is almost as much as in bright sun, and there is some output even in rain.

Two companies that supply this type of solar array are: Solar Power Corporation, 186 Forbes Road, Braintree, Mass. 02184, and Spectrolab, 12484 Gladstone Ave., Sylmar, California 91342.

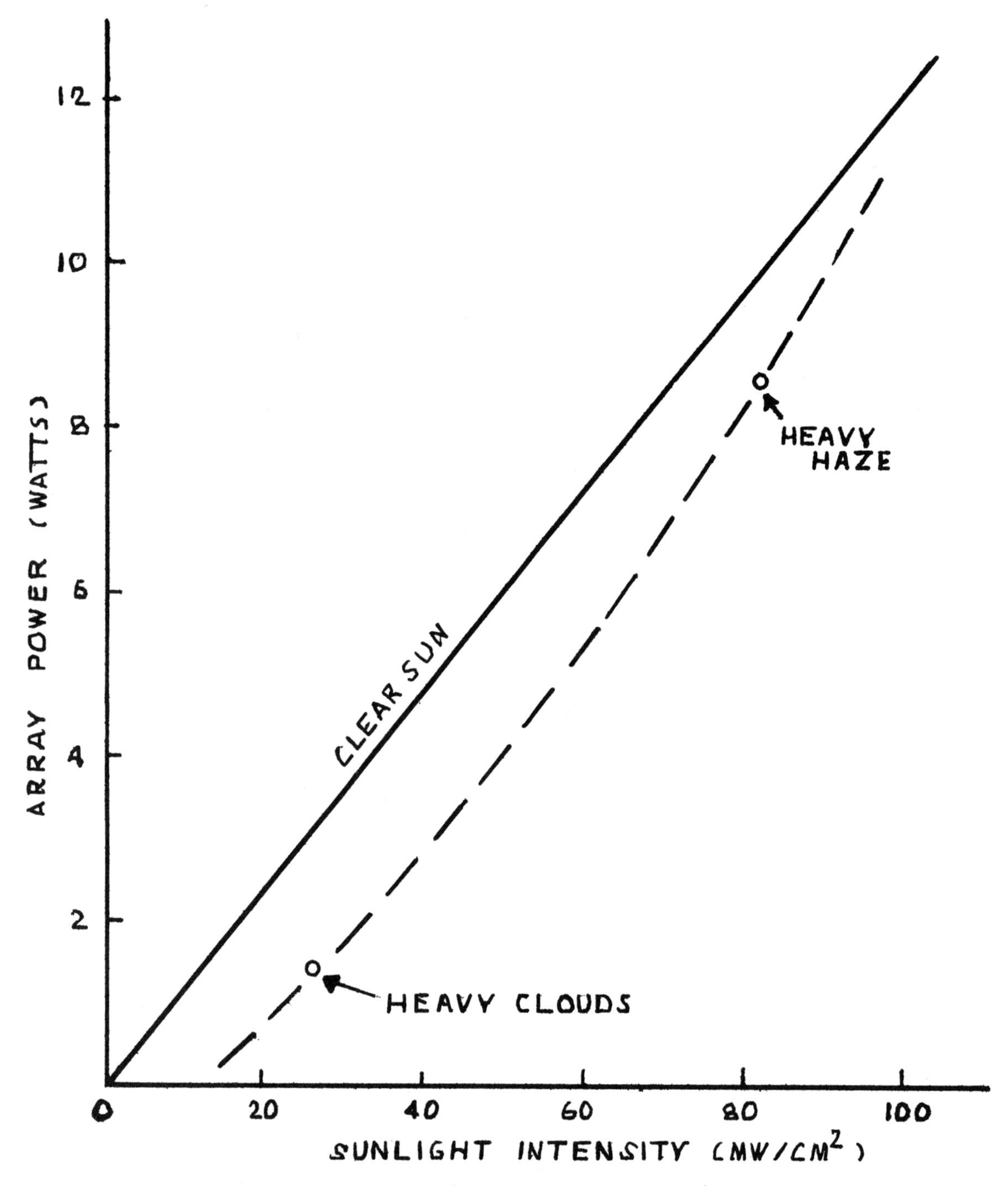

Figure 8-1. Power output of typical 12-volt array as a function of sunlight conditions.

Silicon Ribbon—The Solar Cell Breakthrough

With the cost of silicon wafers running to $2000 per Kw, even highly efficient collectors cannot compete with conventional power—unless someone can find a way to make Beckman's 28 watts/cm^2 work. As was noted above there is not much chance of bringing the price down by improving crystal growth and slicing techniques. Another way would be to find a different method of making the thin wafers.

The second way has great promise. A. I. Mlavsky of Tyco Laboratories, Inc. (Waltham, Mass. 02154), in collaboration with B. Chalmers of Harvard University, has developed a process for producing silicon ribbons of the desired thickness in a continous strip at the rate of an inch a minute. To obtain solar cells the ribbon is simply cut into pieces of the desired length. There is no waste as there is with the crystal boule. The rectangular cell thus obtained has a further advantage. Since the circular shape of the wafer as cut from the boule makes it impossible to achieve a compact array design, the wafer is often trimmed into a rectangle, entailing a further waste of expensive material. The ribbon crystal is already a rectangle without waste.

The process developed at Tyco consists of melting polycrystalline silicon in a quartz crucible and drawing out a ribbon through a die of the desired dimensions. The die conducts the melt to its top surface by capillary action and determines the shape of the solidifying material through surface tension forces. Called "Edge-Defined Film-Fed Growth," the process is capable of fast crystal growth as long as the die has at least one thin dimension. The crystal growth takes place at a point sufficiently remote from the bulk melt that replenishment of the supply is possible by direct addition of raw material.

As the crystal is determined by the die, many dies can be applied to the same melt, and many ribbons can be drawn at the same time. This feature readily leads to mass production.

Since the thin liquid film from which the crystal is grown is "clamped" to the top of the die by surface tension, the growth process is relatively insensitive to changes in pulling rate and crucible temperature. This self-stabilizing aspect of the technique is of particular importance in the development of industrial machinery that can operate over long periods without the need for sensitive and accurate controls.

The chemical purity of silicon for solar cells must be strictly controlled both with respect to the specific exclusion of a family of elements and to the inclusion of certain other specific elements. The elements included or added, in a few parts per million, are those that form "p-n junctions" in the silicon. In semi-conductor parlance, these junctions permit negative electrons to move to positive "holes" when energy is applied and thus to cause current to flow. These purity requirements place a stringent demand on the die material. It must not contaminate the silicon as the ribbon is pulled through. Tyco has successfully solved this important technical problem.

Tyco began working on this new process of crystal growth in 1965. Originally it was applied to the single crystal, aluminum oxide, or sapphire. The process is now being used by Corning Glass Works in the production of sapphire tubes for sodium vapor lamps. Corning draws 20 sapphire tubes at a time from the crucible. Tyco has also licensed RCA and Kyoto Ceramic Co. of Japan to use the process. Thus, the technique has been proven in mass production.

Tyco has already produced enough silicon ribbon by the process to know that it makes solar cells of good efficiency. What remains to be done is to work out some of the bugs attending mass production and to design and build the production machinery and building.

20 megawatts a year

Toward this end Tyco has joined with Mobil Oil to form a new company, "Mobil-Tyco," to mass produce these silicon ribbons. Mobile has put $80 million into the venture. The plant to be built will probably house 12 machines each capable of making 20 ribbons continuously. The 12-unit size seems to be the least size for efficient operation. Mlavsky estimates that these 12 machines will produce 20 megawatts of electrical capacity a year.

Ribbon Cells Are Cost Attractive

Mlavsky estimates that he can make the ribbon at a manufacturing cost of about $15 a pound. If the cost of polycrystalline silicon can be brought down from its present $30 a pound to about $10 a pound—and he thinks it can with adequate research and effort—then the cell wafers will cost about $25 a pound. Adding about $15 a pound for encapsulation and packaging brings the cost up to $40 a pound, or $200 a kilowatt. Doubling this price as the cost of staying in business brings the cost to $400 a kilowatt, a fairly competititve price for electrical capacity. The cost of installation must be added, but when one compares the building of a solar cell farm with a fission plant the alternative is worth considering.

SECTION 2. CADMIUM SULFIDE SOLAR CELLS

How To Make a Cadmium Sulfide Cell

This solar cell, usually called the "cadmium sulfide cell," is actually a sandwich of cadmium sulfide doped with n-type material and copper sulfide doped with p-type material. The interface between the two materials is a p-n junction. If electrodes are applied to the outside of these layers and connected through an external circuit, current will flow when light strikes the copper sulfide surface.

The n-type cadmium sulfide is evaporated to 20 μm thickness onto a zinc-plated copper foil, which serves as the negative electrode. The p-type copper sulfide is prepared on top of the cadmium sulfide to a thickness of 0.2 μm by ion exchange. The copper sulfide surface is covered with a photo-etched gold-plated copper grid, which is the positive electrode. Wider grid spacing allows greater current flow because there is less obstruction to light, but resistance of the copper sulfide limits the maximum spacing. The cell is sealed by cementing on a sheet of ultra-violet resistant Mylar. The cells as presently manufactured have an area of 55 square centimeters. At maximum peak power a good cell delivers 19 ma/cm^2 with 7% efficency.

Practical cells

The practical cadmium sulfide cell was developed by the Institute of Energy Conversion, University of Delaware, Dr. K. W. Boer, director. Before being developed by the Institute the cadmium sulfide cell had been plagued by inefficiencies, instabilities and degradation in use.

The instabilities and degradations seem to be caused by the presence of humid oxygen and copper diffusion into the cadmium sulfide. The humid oxygen fault is reversible by treatment in hydrogen. The problem is avoided in present array panels by circulating dry nitrogen over the cells. The copper diffusion problem limits the life expectancy of the cell to about 15 years. Experiments suggest that proper doping can counteract the effect of copper diffusion and extend the life.

Cadmium sulfide arrays are presently being evaluated on the University of Delaware demonstration house, "Solar One."

Going Commercial

A company, SES, Incorporated (70 South Chapel St., Newark, Delaware 19711), has been formed with Shell Oil Company support to commercialize the cadmium sulfide cell. This company expects to be selling production cells by the end of 1975.

The maximum efficiency of this type of cell is a little over 8%. The present production cells have an efficiency of 5%.

Boer's predicted factory selling price is $15 per square meter. SES expects that it will be 5 to 10 years before the price of these cells can drop to 60¢ to $1.00 a watt where it will be attractive for large scale power generation. As the very thin dimension of cadmium sulfide requires less material than that required for the silicon cell, the cadmium sulfide cell is expected to sell for less.

Figures are not available for the projected production capacity in the coming years, but from the description of the process it seems reasonable that the capacity could be increased quite easily to meet any demand. The amount of production will be limited to some extent by its competitive position with silicon. Cadmium sulfide may be cheaper than silicon, but cadmium sulfide may have a shorter life. Each will probably find a niche where it is best suited. In any case, competition between the two should increase the total use of solar cells. How much increase will result is anybody's guess, but for estimating purposes let it be assumed that the capacity schedule for silicon listed above will be increased by 20% for a total national schedule of solar electrical capacity.

As with the silicon cell, the manufacture of cadmium sulfide cells is not capital intensive.

Thin Film Research

A cadmium stannate film being developed by American Cyanamid is electrically conductive as well as infrared reflective. The present cadmium sulfide cell has a solid metal sheet backwall as the negative electrode and is therefore opaque to visible light. The possibility has occurred to researchers that this film could be substituted for the metal foil backwall of the cadmium sulfide solar cell. The infrared reflectivity might also be used for the generation of heat. The Institute of Energy Conversion is working with American Cyanamid on this project.

SECTION 3. THE GALLIUM ARSENIDE PHOTOVOLTAIC EFFECT

Another Type of Solar Cell

Gallium arsenide is known to have photovoltaic properties, but the efficiency is less than silicon. For space applications silicon has been the preferred material. Recently there has been a renewed interest in gallium arsenide because it has been discovered that the technique of alloy heterojunctions has increased the efficiency to the order of 18%.

The new solar cell consists of an n-type GaAs substrate, a p-type GaAs surface region of the order of a micron thick, and a p-type GaAlAs alloy layer on the p-GaAs surface. The alloy is heavily doped with zinc. As the alloy layer grows, zinc diffuses into the GaAs, forming a p-region. These new gallium arsenide cells have higher efficiencies for three reasons. First, the process permits reliable formation of a p-n junction within a micron or so of the surface of the gallium arsenide. A shallow junction is desirable because all of the radiation is absorbed within a couple of microns of the surface. Second, the overgrowth of the gallium aluminum arsenide practically eliminates surface states on the GaAs that would normally provide fast recombination sites for electron-hole pairs before they can be separated by the p-n junction to provide electrical power. Such recombination sites cause the pairs to be lost and have been the chief cause of low efficiency in the past. Third, the alloy overgrowth makes a good electrical contact with gallium arsenide and has very low sheet resistance. Since the resistive losses in the cell are reduced, the metal electrodes on the surface may be spaced more widely, opening more surface to light.

These cells typically have an open circuit voltage of 0.97 to 1 volt and short circuit current of 18 to 21 ma/cm^2.

The advantages claimed for the new cells are: The alloy overgrowth provides good shielding against proton and electron radiation; much less shielding is necessary to prevent radiation damage than is required for silicon cells. The new cells provide useful output at higher temperatures; for example, at 300 degrees C gallium arsenide produces at least 5% efficiency, while the silicon output would be zero. These advantages would make them well suited for applications in space.

For terrestrial applications, it has been suggested that gallium arsenide could be grown on germanium wafers. Commercial development of gallium arsenide solar cells is not presently active.

Much of the research on this new type of solar cell has been done by H. J. Hovel and J. M. Woodall of IBM Thomas J. Watson Research Center (Yorktown Heights, New York 10598).

New Capital for the Utilities

The solar cell business is not capital intensive. It will require about $3 million for the machinery for Tyco's twelve 20-unit machines, and 5000 square feet to house them. Assuming that the cost would be linear to install enough plant to make 1000 megawatts of capacity a year, the machinery cost would be $150 million.

It would make an interesting comparison if someone were to figure the capital investment of all the thousands of big and little businesses required to make one 1000 megawatt nuclear plant. It would also make an interesting comparison if someone were to calculate how much capital it takes to produce 1000 megawatts of oil or coal. While the calculation may be like comparing apples and oranges, it is likely that the capital needs of solar cell electricity would be much less. The solar cell business would have to grow to 40 times $150 million to supply the needed new capacity each year, but $6 billion is no more than the utilities are willing to spend for a single off-shore floating nuclear plant.

The Federal Power Commission reports that the electric power industry will spend $650 billion between now and 1990 to expand capacity and maintain service. Gambs and Rauth ("The Energy Crisis," Chemical Engineering, May 1971) estimate that it will require $1.4 trillion for new capacity, transmission and distribution during the next 30 years. Where will the capital come from? If the generator industries and the transmission industries and the oil industries and the coal industries are all capital intensive, the mad scramble for capital may become worse instead of better.

But note: if the low-capital solar cell industry began to fill more and more of the needed electrical capacity, the capital needs of the generator, transmission, oil and coal industries would become less and less, and that capital will then become available for the utilities, and others needing capital.

Solar Electricity, When?

Mlavsky thinks it will take 7 to 10 years to get the first production line in operation. He is not optimistic that intensified effort could push the development much faster. However, it does seem that an all-out drive could chop off at least two years from this schedule. It seems reasonable to suppose that 40 megawatts capacity could be in operation by 1985. The following schedule of increase in plant capacity seems reasonable:

Year	Generating Capacity, Megawatts
1985	40
1990	500
2000	2,000
2010	10,000
2020	30,000

By 2020 the rate of installing new solar electricity capacity could be almost double the present rate of fossil steam installation.

Increase the Output with Concentrators

The concentrator idea advanced for flat plate collectors may also be applied to solar cell arrays. That is, collect a lot of sunshine and apply it to less silicon. That way the output is the same for less cost.

E. L. Ralph of Spectrolab, at the annual meeting of the Solar Energy Society in March, 1965, showed that by placing a silicon cell at the bottom of a cone and pointing the large end of the cone at the sun the power output of the cell could be increased by 2.5 times. By opening this cone into a grooved collector a trapezoidal concentrating collector is obtained. The silicon wafers are mounted on the base instead of a heat absorber.

W. A. Beckman (Solar Energy, Vol. 10, 1966) has shown that silicon cells can be operated at a solar flux of 28 watts/cm^2, which is almost 300 times the normal noonday solar insolation. With this level of intensity the power output is 1.5 watts/cm^2, which is 125 times the unconcentrated output of the cell. If this concentration ratio could be achieved commercially, the cost of a solar array using the space-type silicon wafer would come down to about $150 per Kw.

Spectrolab is supplying a commercial trapezoidal groove solar cell collector. The concentrator channel is formed from a commercially available aluminum sheet with an anodized protective coating. The channel is bent to an angle of 30 degrees, and the base of the channel is the width of the solar cell. The ratio of height to width is 2.7, which collects 82% of the insolation. With this loss and the 7% transmission loss of the cover sheet, the effective concentration ratio is 2.05.

Spectrolab is now designing a concentrator with a 10 to 1 ratio that will be capable of delivering both heat and electricity.

Since the power output of the solar array drops as the temperature increases, the logical extension is to cool the cells by drawing off the heat and using the heat as though it came from a flat plate collector.

Silicon is a curtain at the visible spectrum and a window at the infrared spectrum. Silicon is therefore a selective filter that performs a desired function for flat plate collectors. If the solar cells were mounted in a concentrator that doubled the output, Tyco's cost of $400 per kilowatt would be cut in half. If, in addition, the selective filter properties of the silicon were combined in a concentrator so as to reap a harvest of heat along with the electricity, a very desirable power source would become available. Tyco has been exploring these ideas with the University of Houston.

Questions About Rating Solar Cells

The theoretical limit of silicon efficiency is above 18%, but it is not likely that the efficiency will go much above 15% in practical applications. The practical limit of gallium arsenide alloy is not known.

Utility engineers are accustomed to think of energy conversion of at least 40%. Conversion at 5% to 15% hardly seems worth the bother.

However, there is a different way of looking at solar cell efficiency. No one knows for sure whether the conventional way of rating actually measures the input-output ratio of a solar cell. Conventionally the power out of a solar cell is compared with the caloric measurement of a pyranometer. It is a good question whether the heat absorption as measured in a pyranometer bears any relation to the electricity produced in a solar cell.

Some semi-conductor physicists estimate that 50% of the photons entering some solar cells succeed in causing an electron flow. If such is the case, the efficiency ought to be 50%. The true efficiency measure should be the ratio of the number of photons entering to the number of electrons emitted. Hovel, Woodal and Howard at IBM have plotted short circuit current per photon as a function of photon energy in electron volts, a truer measure of solar cell capacity, but this practice is not common, nor are there practical instruments for measuring this kind of response in the field.

Solar cells show an interesting anomaly when the efficiencies are measured conventionally under different air mass levels. Air mass is a term that quantifies the optical absorption and scattering in the earth's atmosphere, where air mass 0 (AM0) is in space and air mass 1 (AM1) is at the earth's surface on a clear day. On a hazy day it may be AM2. It happens that when the efficiency of a solar cell at AM2 is measured it usually shows a greater efficiency by a couple of per cent than at air mass 1. Since there is greater absorption and scattering at AM2, it seems that the cell efficiency should be less.

The photon seems to behave differently when its energy is re-radiated at a different wave length than when its energy is converted to electron motion.

However, the conventional expression of efficiency for a solar cell is acceptable if it is taken as a rating factor, or value factor. An efficiency calculation rates one cell against another, or against a standard cell, or places it in a family of cells. The pyranometer measurement indicates the insolation or solar intensity at a given moment. Insolation is a relative value. The solar cell's relation to relative insolation is a relative value.

Efficiency is a convenient way to rate solar cells, and this Report will continue to follow the conventional practice.

Other factors affecting efficiency

The efficiency of a solar cell is affected just because it is part of an electric circuit. The best efficiency of a power source is attained if the load and the power source are matched at the maximum power point. However, the maximum power point varies constantly during the day. To maintain best efficiency the load would have to be varied all the time, an impractical solution. (For a valuable discussion on the selection of optimum resistance for a solar cell circuit see the Rao and Padmanabhan paper in Solar Energy, Vol. 15, pp. 171-177.)

The efficiency of a solar cell is affected by temperature because an increase in temperature causes a voltage drop.

The Model Community's Solar Electric Resources

Assume that all of the area previously occupied by flat plate collectors now is filled with solar cell arrays. What would be the output capacity?

Recall that the flat plate collectors were calculated to produce 5.455 trillion BTU per year. This quantity, at the conversion rate of 3413 BTU per Kwhr, is 1.6 billion Kwhr. But remember that the flat plate collector figures used 50% efficiency values. The 100% electrical equivalent value is therefore 3.2 billion Kwhr.

Staying with the convention that the solar cell output can be computed as an efficiency rating of the maximum value, and assuming that the solar cell efficiency is 10% of the total insolation, the community could produce 320 million Kwhr. Using Edison Electric Institute averages for the year 1973, the total electrical needs of this community are about 86 million Kwhr. The collector resources of this community therefore have a potential surplus of 234 million Kwhr.

It has been shown to be practical to combine both heat collection and electric generation in the same collector. It seems reasonable to expect that third or fourth generation combined collectors will produce heat at 70% efficiency and electricity at 15% efficiency. Any system capable of energy conversion at 85% efficiency is worth taking another look at. With efficiencies of that order a great many communities should have no great trouble being self-sufficient in energy consumption.

9

WIND AS AN ENERGY RESOURCE

HOW MUCH WIND ENERGY IS THERE?

The amount of energy developed in a hurricane or a tornado is greater than that developed in a hydrogen bomb explosion. Even a gale force wind must be respected. Man-made devices have to retrench in the forces of great winds. There is no present technology capable of converting the energy of great winds. Those man-made conversion devices, the windmills, are constrained to operate in a narrow band of about 8 to 60 miles an hour.

It is much more difficult to estimate the amount of recoverable wind energy than the amount of direct solar conversion. Some of the difficulties are:

—The recoverable wind energy is a small fraction of the total amount, but no one knows the whole number. With direct solar energy there is a known maximum; the recoverable amount is a known percentage of the maximum.
—Wind energy is randomly intermittent. There are periods of calm interspersed with periods of flow from 4 to 8 mph and with other periods much higher than 8 mph. Elaborate steps have to be taken to obtain predictable statistical averages.
—The ultimate limit to the recovery of direct solar energy is number of square feet of the earth's surface. The ultimate limit of wind energy square feet is not measureable because the surfaces normal to the wind are in the vertical direction. How high is up? That question is meaningful to the wind engineer because wind velocity and constancy increase with height.

In a sense the wind is one of nature's solar energy storage systems. The sun heats the earth's surface causing the air to expand and to build up a pressure gradient between one region and another. The pressure gradient is the storage system, and the wind is the conversion device for relieving the pressure. All man can do is stick up a few gadgets here and there to catch some of the wind as it goes by. He can never hope to control the whole pressure storage system and therefore has no true way to answer the question, How much wind energy is there?

The only practical question to ask is, How much wind energy can man harvest at one locality? That question is answered by combining responses from two sources: 1, a computation of the output capacity of the wind machine; and 2, long term measurements of wind speeds at a chosen locality.

How to Compute the Recoverable Wind Energy

Computation of wind power derives in the ultimate from the fundamental physical equation for kinetic energy,

$$KE = \frac{1}{2} m v^2$$

where m is the mass of the air molecules
and v is the velocity of the air molecules, or wind speed.

The power (energy/time) expression is,

$$P = \frac{1}{2} p A v v^2 = \frac{1}{2} p A v^3$$

where A is an area normal to the wind direction
and p is a density function of pressure, temperature and humidity.

Vaughn Nelson's report "Potential for Wind Generated Power in Texas" for the Governor's Energy Advisory Council (Dept. of Physics, West Texas State University, Canyon, Texas 79105) contains an excellent derivation and description of the method of calculating wind power.

A common expression for calculating wind power is,

$$P = K D^2 v^3$$

where P is expressed in suitable units such as Kw
K is 0.593, a derating factor representing the commonly accepted density function
D is the diameter of the rotor
v is the wind speed.

That is, power out of a windmill is directly proportional to the area of the capture surface and to the cube of the wind speed multiplied by the derating factor 0.593.

The recoverable wind power is further limited by the efficiency of the windmill. The blade efficiency of a rotor is something in the order of 70% at the theoretical limit. With the wind energy already brought down to 0.593, a further reduction of 70% brings the windmill efficiency down to 41.5%. With losses from bearings, vibration, etc., wind engineers consider the maximum practical efficiency to be about 35%. The practical power equation ought to be: $P = 0.35 D^2 v^3$.

There is a further limitation, related to generator design, that is as restrictive as efficiency. Engineers will appreciate that practical generators operate only at a specific design speed, or at most over a narrow speed range. Designers have the option of a generator with a variable but narrow speed range or a constant speed generator with rotor speed controlled by variable pitch blades. In either case there is a narrow wind speed range below which and above which no energy is taken. That is, the velocity above and below the range is wasted. Windmill efficiency cannot be reasonably expressed as a ratio of power out to available power in because power available and recoverable power in are different. Windmill efficiency ratings refer only to the design velocity range.

Somebody Should Go Wind Prospecting

E. W. Goulding, English wind power pioneer, as long ago as 1960 was insisting on the absolute necessity for taking a long-term wind survey of a site before installing a windmill. The dependence of wind power on wind velocity cubed is the reason. A difference of even 2 mph in average wind speed can make a significant difference in the total output for the year. Differences of 2 to 5 mph can easily occur within a locality.

Wind data should be taken for at least a year at a site, and longer if possible. Data for less than a year are of little value.

Wind data are of two types, those taken for a particular site and those taken for a region. Regional data are marginally acceptable for site evaluation. Regional data are best used to determine whether it would be profitable to test sites within the locality.

Somebody should be prospecting for good wind power regions. Vaughn Nelson for Texas and Tunis Wentink, Jr. for Alaska have made comprehensive surveys. E. Wendell Hewson is currently doing a survey and study of Oregon wind resources for the Oregon Peoples Utility District Directors' Association.

Nelson and Wentink used Weather Bureau and airport data taken with conventional anemometers because these information sources are the only ones available. These sources are inadequate, however, for wind power prospecting. Weather Bureau anemometers are placed either at locations representative of a general area or at stations chosen largely for temperature and solar measurements or general weather conditions. Airport anemometers are usually placed at a height of 13 feet because pilots want ground speed information. Windmill designers need wind speed data from much higher elevations. If an anemometer happens to be placed too close to a building the reading could be 2 or 3 mph faster or slower than the true speed. Data intervals from these sources are also too wide.

The government is belatedly spending a great deal on research for non-fossil resources, but one of the weakest aspects is the lack of accumulation of basic data. Detailed wind information is even more important for wind than for sun because there are more variables.

A sincere national commitment to wind energy would include a program for wind resource prospecting. Such a program would include the siting of several hundred automatic wind recording stations throughout the country. All regions would be sampled so that the possibilities of all parts of the country could be known. It would be desirable to have some instruments at 1500 feet and many more at 150 and 50 feet. It is likely that TV towers and some tall buildings could provide the mounting for these instruments without having to build special towers.

The technology and manufacturing drawings for these automatic wind recording instruments already exists. The drawings and engineering know-how exist at Geodyne Corporation, now the Environmental Equipment Division, EG&G (151 Bear Hill Road, Waltham, Mass. 02154). A program of acquiring oceanographic data, including wind speed and direction, has been going on for several years. The research data and technological know-how can be found at the Woods Hold Oceanographic Institute, the National Oceanic and Atmospheric Administration, and probably other oceanographic institutes.

These instruments measure wind speed and direction in digital format and either record the data on magnetic tape or telemeter them to a central station. Being in digital format, the data are easily processed by computer. Data now available from weather stations and airports are taken at one to three hour intervals. Automatic instruments can take and record readings at as short an interval as desired, even down to seconds. These short interval data would provide detailed analysis of gusts and swirls—very valuable information to the windmill designer.

The wind data program recommended here would convert its raw data into several different formats. One of the most valuable formats for general information would be a series of contour maps. These maps would show equal energy contour lines for average wind at different heights, and the probability of various wind speeds at different heights (e.g., the probability that the wind will be 8 mph or higher and at 15 mph or higher at different heights).

Boiling It Down to Site Selection

There are isovent maps available from meteorological information sources. Although these maps are not as valuable as the equal energy contour maps described above, they do help in deciding whether a region is worth prospecting for specific sites.

Sites must be found that promise an economical yield of Kwhr per installed Kw. A. H. Stodhart (Electrical Research Council, Surrey, England) argues that at 1973 fuel prices sites must be found having mean wind speeds of about 30 mph for long enough times to produce at least 3500 Kwhr per year per installed Kw. Otherwise an economic case cannot be made for the windmill. If Stodhart's criterion were to guide the decision to build windmills, not many windmills would be built because not many sites can provide that much 30 mph wind. His opinion does show what must be uppermost in mind for site selection: where does one find a large amount of sustained wind?

A consortium might contract to have a region surveyed for site selection. Surveying for non-fossil resources, however, is a profession with no present standing, and probably no present commercial surveyors exist. An alternative might be a geophysical department of a university. Hewson's Oregon study referred to above is a case of such a contract. Hewson's group has identified 150 possible sites along the Oregon coast and Columbia River.

Surveyors need to look for promising sites instead of random locations. There are some guide lines. For example, smooth shaped hills with all-around exposure provide mean speeds of 35% to 50% above lowlands in the same area.

Conical shaped hills without flat tops are preferred to ridges. The best slope characteristic is about 16 degrees in the last 1000 feet from the summit.

Hills offer another advantage other than creating higher mean speeds. In level country of moderate roughness the wind increases with height at some rather constant rate. For large windmills this variation produces a significant difference between the tip of the blade at its highest point and the tip of the opposite member near the ground. This difference might be as much as 5 mph. Considering that the power of the wind varies with the cube of the speed, this difference can greatly influence the stresses and fatigue factors placed on the rotor and tower. If the wind sweeps up a hillside, this vertical gradient becomes compressed. For example, if the gradient over level ground amounted to 5 mph difference through the first 200 feet, the difference in the same height at the top of the hill might be only 1 mph.

If there are hills and valleys, some places may form a vortex where wind speeds increase. It is even possible to do some contour shaping of hills and valleys to intensify the vortex effect.

Accessibility to the site is an important criterion because installation is one of the larger cost factors of windmills.

WHAT SITE SURVEY DATA ARE IMPORTANT?

The raw data from the survey instruments will be wind speed and direction bytes at intervals as frequent as practical. Processed data will be required by the engineer. Each engineer will probably have his own idea of what data should be processed out, but certainly he will want the following: The mean speed for each month, the fastest speed for each month and the annual mean speed and fastest speed for the year. The other processed data he will want will depend on the kind of curves he wants to draw for the site.

One of the valuable curves is the wind speed histogram. The X-axis is in wind speed, and the Y-axis is in number of observations at the given wind speed. Thus, if there were 300 observations, or times, when the wind reading was 3 mph, 600 times when it was 5 mph, 250 times when it was 8 mph and 50 times when it was 10 mph, then the curve will be highly peaked at 5 mph. Almost all of the wind energy will be at 5 mph. The computer printout will need to report the number of readings taken at each selected wind speed.

Another valuable curve is the annual available power. The X-axis is wind speed, and the Y-axis is Kwhr in square feet per year at the speed indicated. To insure that the curve represents available power the Y-axis must be multiplied by 0.593. The computer will be asked to figure the plotting points by computing the Kwhr per year per square foot available at a range of wind speeds. The computer will obtain the amount of time spent at each selected wind speed from the raw data. A typical available power curve is in Figure 9-1. This curve represents graphically and quickly the best windspeeds for the most power at the site and the bandwidth of winds over which the most power can be drawn. The integration under the curve represents the total available wind power per square foot at the site, although a faster way to determine this amount would be to ask the computer to sum it up after deriving the plotting points. This curve tells a designer almost at a glance how large a windmill he will have to provide in order to obtain a given total power.

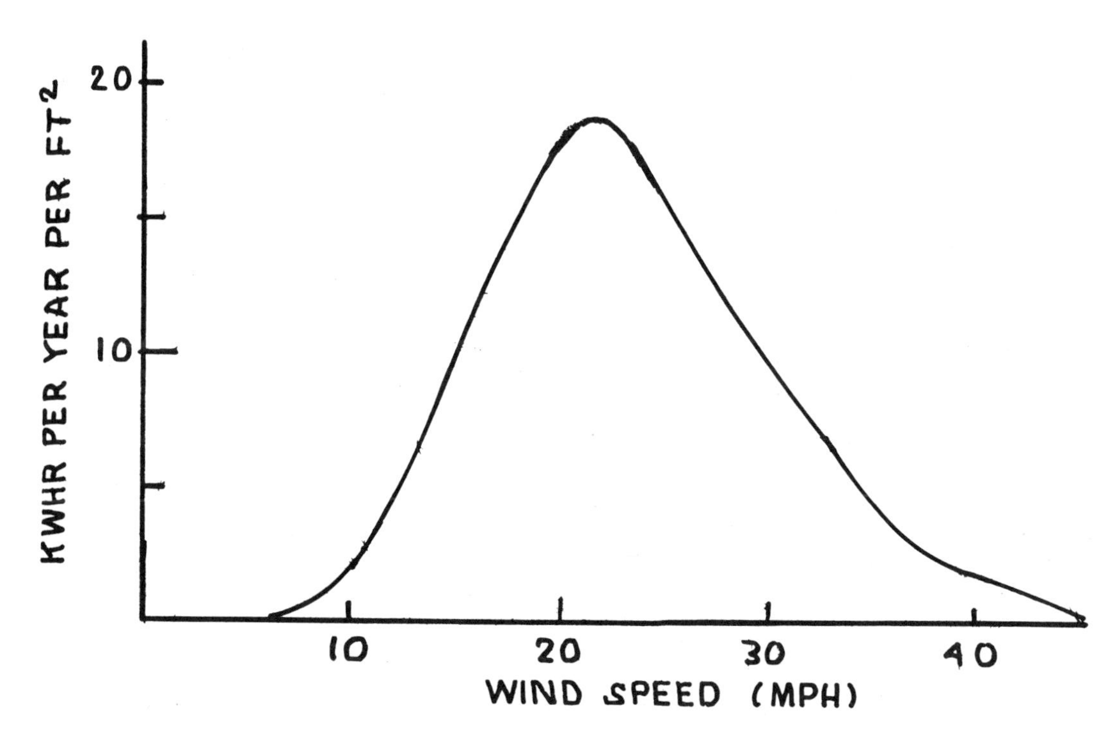

Figure 9-1. A typical annual available power curve.

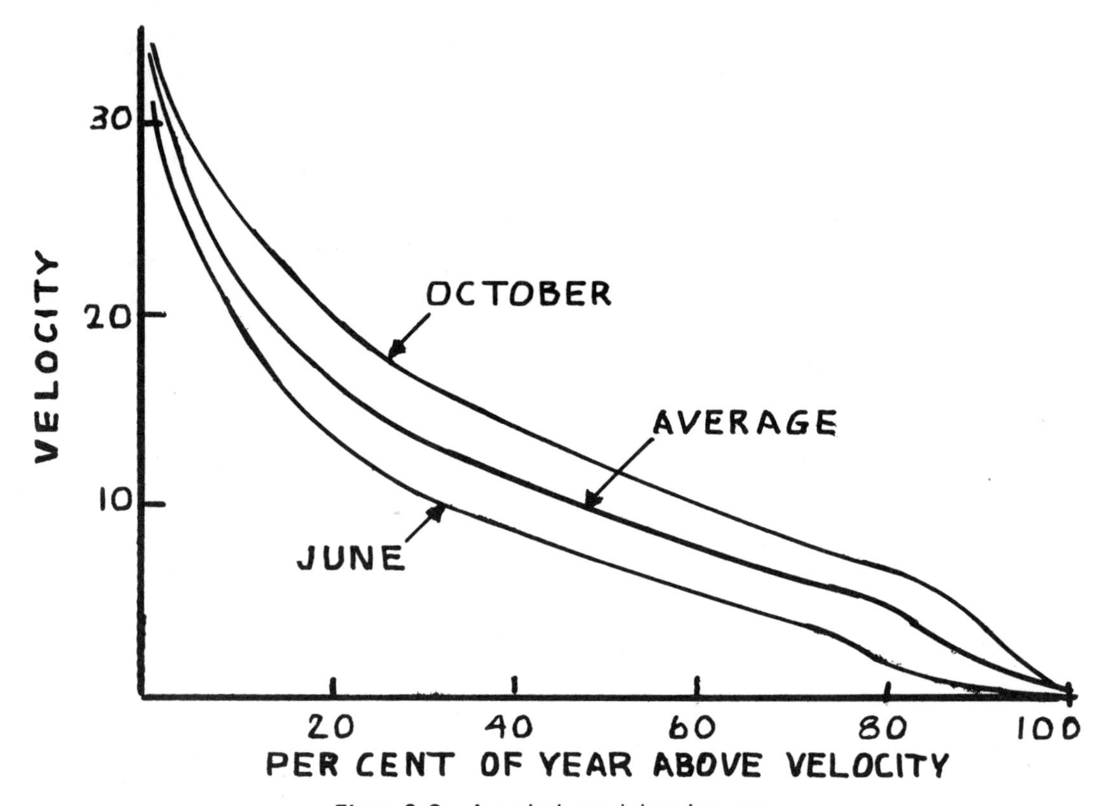

Figure 9-2. A typical speed duration curve.

The speed duration curve is the most widely used and probably the most useful type. The X-axis is in per cent of time above the indicated speed, and the Y-axis is in wind speed. The method of determining the points on the grid is as follows: First set up a list of speed ranges having a narrow spread such as 3 mph, which is equivalent to the Beaufort force number scale. For example, 0, 1-3, 4-6, 7-10, and so on, mph. Next determine the number of readings taken within each of the speed bands and tabulate against the list of speed ranges. Then calculate the percentage each of these numbers represents of the total number of readings. These percentages are the X-axis points of a curve where the Y-axis is in wind speed.

The speed duration curve enables the designer to see at a glance the amount of time during a year when the wind speed will be at or above an indicated speed. For example, if the design speed of a windmill is 20 mph and the curve shows that 20 mph occurs 25% of the time, it is quickly apparent that the machine will be producing rated output 2190 hours a year. A typical speed duration curve is in Figure 9-2.

Tunis Wentink, Jr., Geophysical Institute of the University of Alaska, has done some analysis of the speed duration curve. (Refer to Scientific Report: "Wind Power Potential of Alaska, Part I, Surface Data from Specific Coastal Sites" prepared for the National Science Foundation.) His analysis shows that it should be possible to predict a speed duration curve from the mean wind speed of short periods such as a month. He indicates that the computed curve is reliable over the range between cut-in speed and maximum speed before feathering. If the Wentink method works out in practice it should reduce the cost of prospecting by 91%.

The designer needs to know the wind speed at the hub if he is designing an axial flow windmill. Lacking survey data at his design height, there are equations allowing one to extrapolate from data taken at lower height. (Refer to O. G. Sutton, "Micrometeorology," McGraw-Hill, 1953, and R. E. Munn, "Descriptive Micrometeorology," Academic Press, 1966.) Some calculations use the gradient factor G^p, where p is about 1.4. Thus, if the ground level wind is 20 mph, at some height above ground it will be a speed such that the gradient is $G^{1.4}$.

10

HOW ENGINEERS DESIGN WINDMILLS

CLASSES OF WINDMILLS

There are two general classes of windmills. One is the small size usually running from 1 to 10 Kw capacity, although some may go as high as 30 Kw capacity. There is no hard division line. These windmills are mass produced and of general design. At present the most common application is with an individual who has a ranch or a home too far from the power line. The other class has a rating of 100 Kw or larger, although again there is no hard division line. Only two or three windmills larger than 100 Kw have ever been built, but NASA Lewis Laboratory (Cleveland), the main arm of the government for large windmill research, has plans for building 1 megawatt and perhaps larger sizes. The large windmill will almost surely be custom designed for the site.

The small windmill designer needs to have the data gathered by the site surveyors and wind prospectors because he needs general information about wind behavior. However, the custom designer of the large windmill will find site survey data of crucial importance.

WHAT ARE THE IMPORTANT DESIGN CONSIDERATIONS?

The cost of a unit of windmill power is the cost of design, manufacture, installation and maintenance amortized over the life of the machine, say 20 years, divided by the power produced.

Traditionally, a power generating device has been judged on the basis of its ability to deliver power competitively with conventional generators. Windmills have fared poorly in this comparison by about 3 to 1. There is some question whether this comparison is the proper way to rate a windmill. For example, Joseph Tompkin, a consulting engineer in Salem, Oregon, argues that the operating cost of a wind power plant is not a function of the installed capacity but instead depends on the number of wind plants operating in an area. The more plants operating in an area the lower the cost. In a feasibility study he did at Cascade Locks, Oregon, he noted that one unit could produce 14 million Kwhr at a cost of 4.3 mills per Kwhr, while if 20 units were operated in the same area the cost per Kwhr was 0.8 mill.

Whatever cost criterion is used it is still important to keep the cost of an individual unit low. As Louis Vadot, British pioneer E. W. Golding*, and others pointed out long ago, cost is the biggest factor of design. It is the factor that throws out many good ideas.

Rotor Design

Costs have committed most designers from Golding onward to settle on the two-blade rotor for large windmills.

The most famous windmill ever built in this country was the Putnam machine installed at Grandpa's Knob in Vermont in 1941. This windmill had a 175-foot, two-blade rotor. Hewson, in his first report to the Oregon P.U.D. Directors' Association, has compared the probable 1971 costs to the actual 1941 costs of the Putnam windmill. His computations show that the two-blade rotor cost is 10% of the total installed cost. Prorating into a three-blade rotor, the rotor would be 14% of the cost. In contrast, the estimated cost of the rotor for the experimental 100 Kw windmill being built and tested by NASA Lewis Research Center is 46% of the total estimated cost; if the rotor were three-bladed the cost would be 56%. Since modern blades will be much more sophisticated than those on the Putnam machine, the cost will be much higher, and the motive for designing for two blades will be much greater.

The greater the number of blades the greater the starting torque and the lower the threshold wind velocity for starting. However, the greater the number of blades the lower will be efficiency per blade, although the amount of reduction is quite small.

Smaller windmills often use more than two blades. M. L. Jacobs (Jacobs Wind Electric Company, Inc., Fort Myers, Florida 33905), who produced

---

*Golding's classic book, The Generation of Electricity by Wind Power, has long been out of print, but it is now being reissued by its London publisher.

thousands of 15-foot rotor windmills (2500 to 3000 watts) between 1931 and 1957, settled for a 3-blade propellor because he found that there was excessive vibration when the 2-blade type shifted in the wind.

Blade Design

Blade design is the most important factor of windmill efficiency. Blade design is a highly technical field, but it is closely related to aircraft airfoil, propellor and helicopter rotor design. There is plenty of engineering talent and manufacturing skill for the construction of superior blades at reasonable cost. The Putnam windmill mentioned above was abandoned in 1945 because of fatigue failure in one of the blades. The blade was not replaced because of wartime shortages. With today's technology the blade failure would probably not have occurred.

Although a helicopter rotor cannot be turned on its side and operated as a windmill rotor, there is great overlap in the design characteristics of each. A utility manager needs an insight into rotor concepts if he plans to purchase a windmill. A summary of the relations between helicopter and windmill rotors follows.

Successful large, reliable, low-maintenance windmill rotors must minimize the dynamic response to aerodynamic, inertial and gravitational forces. If exciting forces occur at frequencies near the natural frequencies of the rotor (or the tower), resonance may seriously amplify the dynamic loads. The rotor blade must be designed to avoid resonance to achieve low fatigue stress and long life. One difficulty is that during operation below rated speed, or if ungoverned, it is virtually impossible to avoid resonance at some speed.

The elastic deformations of rotor blades include: flap, or vibratory bending of the blade tip in a direction perpendicular to the rotation; lead-lag bending of the blade in a direction parallel to the direction of rotation; blade torsion or elastic twist; vertical and horizontal bending of the rotor shaft should also be included.

Equations exist for determining the rotating natural frequencies at which these vibratory loads occur. (Refer to D. A. Hodges and R. A. Ormiston, "Stability of Elastic Bending and Torsion of Uniform Cantilevered Rotor Blades in Hover," Paper No. 73-405, AIAA, March, 1973.) These loads or forces exist whenever the rotor blade is not uniformly loaded around the azimuth. The mean axial wind component generates thrust forces that affect the blades equally downwind (called "coning" in helicopter parlance). Gradients in axial velocity—the vertical gradient, for example—produce a tilting of the rotor disk with respect to the shaft. Non-uniformities in velocity peculiar to the windmill location, the tower wake, and atmospheric turbulence produce important unsteady loads.

Stresses due to inertial blade loads include centrifugal tension from rotation, lead-lag and flapping loads from Coriolis forces arising from blade oscillations, and gyroscopic forces from precession of the rotor shaft to maintain alignment with the wind velocity vector.

Gravity loads may produce significant lead-lag bending stresses for large rotors. Static deflection from gravity on the rotor at rest could also be a problem. With helicopters, the plane of the rotor is normal to the gravity field. When they start up, centrifugal stiffening occurs uniformly; the wind rotor runs vertically with the result that centrifugal stiffening is uneven.

Helicopter theory can offer some comments about the ultimate size of large rotors. For geometrically similar constructions, which means that a constant level of aerodynamic efficiency at a given wind speed is assumed and the tip speed is a constant value, the important aerodynamic and centrifugal stresses are independent of rotor size. The natural frequencies remain constant, and the resonance characteristics are not influenced by rotor size.

However, aerodynamic efficiency trade-offs will alter the ground rules of geometric similarity and constant tip speed. For example, power losses due to drag can be reduced by reducing tip speed and increasing rotor solidity. (Solidity is a helicopter term referring to the ratio of the blade area to the disc area.) Increased tip speeds would be advantageous for reducing the cost of the gear box between the rotor shaft and the generator.

Although the aerodynamic stresses are independent of size, the weight of the rotor increases by the cube of the size. The gravity effect increases proportionately with weight. The economies of scale do not go on endlessly with windmills because gravity will eventually limit the size.

Giving attention to the selection of materials can reduce the weight and minimize some of the vibration problems. Much work needs to be done yet on the use of glass fibers, epoxies and other plastic materials for both aerodynamic and structural characteristics.

Interrelation of Rotor and Tower

In most regions of the country the maximum wind speed (except tornadoes and hurricanes) exceeds the average wind speed by at least 6 times. The entire windmill, both rotor and support, must be capable of withstanding minor hurricane forces. Depending on the blade pitch angle, the aerodynamic drag of a freely milling rotor can be as much as ten times that of the same rotor locked at zero RPM. Since the rotor is the dominant structural part contributing to the drag, particularly for large windmills, some sort of drag control is essential. In some machines, especially the smaller ones, a braking system, or drag control, may be activated by centrifugal force, wind velocity or power output level. The Jacobs windmill used a flyball-governor-operated variable pitch speed control. The NASA Lewis 125-foot machine achieves constant speed rotation by means of a variable pitch control articulated by the hub. The Princeton "sailwing" rotor uses blade twist to achieve speed control.

The heavy drag force from the rotor causes a large overturning moment on the tower. To the horizontal stress imposed by the rotor must also be added the horizontal stress induced by the tower's own resistance to the wind. The tower must withstand the horizontal forces in all directions, since the rotor will be capable of changing 360 degrees in azimuth to accommodate changes in wind direction.

Changes in wind speed induce vibratory forces on the tower. Hence, the tower materials are subject to fatigue problems.

One of the most serious potential dangers to the windmill is overspeeding of the rotor. Not only would maximum high winds dangerously increase the drag on a free-wheeling rotor, they could also cause the rotor to self-destruct. The structural integrity is dependent not only on the strength of the tower but also on the regulating system. If the regulating system is not foolproof, the tower will be destroyed even if it is 10 times stronger than necessary.

Tower design

The braced structural steel tower is the type most commonly used because there are many manufacturers capable of producing this type of construction. A tubular steel tower has sometimes been used. In some opinions the latter is more aesthetically pleasing.

The cost of the tower increases at a rate greater than the linear increase in height. It is desirable to put the rotor hub as high as possible to take advantage of the increase of wind speed with height. The cost of the tower influences the installed cost of the windmill and therefore the cost per Kw of installed capacity. The windmill designer must harmonize a number of trade-offs to achieve a maximum yield and a least cost per Kw. The NASA Lewis estimate for the cost of their tower is 10% of the total.

It is, of course, easier to put small, rather than big, windmills on the top of high towers, where height is relative to the rotor diameter. A popular tower for small windmills is a telescoping tube which is raised by an interior winch.

Location of rotor

Largely through the influence of the German windmill designer Ulrich Hutter, most modern designers prefer to locate the rotor downwind from the tower. The argument is that if the rotor is located upwind the wake of the passing rotor causes changes in wind forces on the tower, which subjects the tower to pulsating forces. If the rotor is located downwind, the vibrations on the tower are reduced. However, the blades are then subjected to vibratory forces because they pass through the wake of the tower. It is argued that, given a choice, it is less catastrophic for a blade to fail than for a tower to fail. It may be argued otherwise, however, that a catastrophic failure of the blade would almost surely mean a catastrophic failure of the tower.

Ice and snow removal from the rotor

Accumulation of ice and snow on the blades has worried many designers, but experience with windmills has shown this concern to be more fear than fact. The reason, apparently, is that blade vibration prevents these encrustations from accumulating. Some amount of vibration of the blades is necessary in northern climates.

Yaw control

With small windmills up to about 15 feet in diameter yaw control, or the automatic facing of the rotor into the wind, is usually handled by a tail vane. Some models, however, have put the rotor down wind and have depended on the drag on the rotor to produce the turning moment.

Above the 15-foot windmill the gyroscopic forces of the rotor begin to play an important part in yaw control, and the tail vane becomes less satisfactory. Larger windmills depend on auxiliary control instruments or mechanical devices to sense changes in wind direction and speed. The rotor heading is changed by a power driven mechanism.

In the NASA Lewis windmill the rotor shaft and power take-off rests on a large bull wheel at the top of the tower. A versible motor driving the pinion gear responds to wind vane-anemometer control. The rotor is turned into the wind in response to a change in wind direction of 10 degrees whenever the wind speed has been 8 mph or more for 5 minutes continually.

Yaw control must be operative at all times even when the blades are feathered or locked. Yaw control should keep the rotor heading into the wind to within less than ±12 degrees.

Hubs

Large windmills always have variable pitch blades. The mechanical gears, linkages, actuators, etc., necessary to vary the pitch are housed within the hub.

The fixed hub design bolts the hub rigidly onto the main low-speed shaft. This type of hub allows only the pitch change degree of freedom. This design does not allow any articulation that may be required due to forces on the blades from wind shear and tower shadow effect.

An alternative design, the teetered hub, allows articulation of the hub to relieve cyclic moments on the shaft. In this design the low-speed shaft terminates into a tee with the hub under-slung on this tee. This design has been used on helicopters for years.

The pitch change mechanism is usually operated hydraulically.

Transmission Train

Windmill drive shafts turn slowly compared to other machinery. The rotor speed of the NASA Lewis 100 Kw windmill is 40 rpm. If the power converter is an electric generator a gear train must step up the rotation by a substantial ratio. The NASA Lewis gear box has a ratio of 45/1.

HOW TO USE THE WINDMILL SHAFT HORSEPOWER

The most commonly used power converter for the windmill is the electric generator. Within that category there is a family of devices.

Small windmills usually drive a dc generator because the output almost surely charges a storage battery.

The large sizes likely to be of interest to the utilities would use ac generators. In this case, of course, the problem is how to derive 60 cycle current from a variable speed source.

One solution to the variable input problem is to design the windmill to maintain constant shaft rotational speed. The rotor speed is controlled by varying the blade pitch. This solution requires a sophisticated automatic control system to maintain a precise 60 cycle current.

Another solution is the device called the "field modulated frequency down converter." This converter system is built around a high frequency alternator. The field of the alternator is modulated at the desired output frequency. Whatever the shaft speed of the alternator the output current is at the field modulation frequency. If the output were to be coupled into the power line, the field would be modulated by the power line. A 10 Kw model of this converter has been successfully built and tested at Oklahoma State University. (Refer to "A Synopsis of Energy Research," College of Engineering, Oklahoma State University, Stillwater, Oklahoma 74074, or contact R. G. Ramakumar, the same address.)

Since the generator is of the high frequency type it is lighter and smaller than conventional models of the same capacity. This feature is very attractive to windmill installations where the generator has to be mounted at the top of a high tower.

This generator offers interesting possibilities to the utilities for applications other than windmills. For example, high speed turbines could drive the generators directly without gear reduction. The necessity for constant speed control of the generator is eliminated.

Mechanical devices for maintaining a constant speed output with a variable speed input have also been invented.

Researchers at the University of Wisconsin have designed and built a three-phase ac/dc/ac converter capable of accepting variable Hz from a wind generator and delivering either rectified dc or rectified 60 Hz. The ac inverter operates from the dc rectifier. The inverter is the same circuit already used by the power companies for the present day large scale dc power links. With the present state of the art the rectifier-converters can be built to handle up to 3000 megawatts.

Compressed Air

Considerable thought has been applied to the idea of using an air compressor instead of an electric generator for windmill power take-off. The conclusion reached so far is that the most feasible way for a utility to convert windmill compressed air is to store it for peaking and to burn it in a gas generator.

Injection of compressed air into the gas turbine has been shown to be practical, and fuel savings of as much as 66% can be achieved. Since gas turbines are already used for peaking, the compressed air route seems a good one to follow.

Since compressed air in huge volumes would be required, the basic problem is to find storage capacity large enough. Turning to the natural gas industry for storage ideas, this industry routinely stores natural gas in depleted oil fields, aquifiers and solution-mined salt caverns.

A specific example of gas storage may be quoted. In Mississippi two caverns have been solution-mined in a salt dome. The volume of each cavern is 5.7×10^6 cubic feet. Gas can be delivered from the two reservoirs at the rate of 375 scf/day. If compressed air were stored in these caverns instead of gas, the storage investment would be 90¢ per Kwhr at 1000 psi.

Compressed air has been offered as an alternative to pumped hydro storage for peak shaving. In this application the off-peak base load capacity could be used for compressing air, in which case the application of wind power to the job is incidental. In practice, windmills could be coupled directly to the power grid without storage.

Windturbine System, Inc. (53 Liberty St., Walton, N.Y. 13856) uses a windmill to compress air, which drives a generator. The exhaust from the generator is used as a refrigerant.

11

WINDMILL TYPES

Most of the development money in this country is going into the large, axial flow windmill. The aim is to optimize the 100 Kw size and then to go to the megawatt or larger sizes. There are, however, many other lines of experiment, invention and development going on. Although they are not as richly funded as the large size, some of them are quite significant.

THE SMALL HORIZONTAL AXIS WINDMILL

A substantial foreign manufacturing business exists for the small electric windmill. Over the past 40 years many thousands of these have been built. They are successfully used, however, only at locations where utility power is not available. The problem is cost. They produce power at about 10¢ a Kwhr.

It would be quite expensive for a home to become completely dependent on these small model windmills. It would take about 10 Kw installed capacity to handle peak demands during long periods of calm. Sufficient battery capacity would have to be provided to handle this amount of storage. The batteries could be coupled in series-parallel to deliver 110 volts dc for lights and stoves, but inverters would have to be bought to operate the TV and motor-driven appliances.

Special long-life batteries for windmill home storage have been developed and marketed.

During the '60s and early '70s the only commercially manufactured windmills were those made in France, Switzerland and Australia. It is possible to buy from sales agents in this country (e.g., Solar Wind Company, East Holden, Maine 04429), but one encounters the usual problems of high import duty and delay in shipment. Some trickle of windmill products are beginning to appear with American firm names. By 1980 it will probably be possible to buy fully developed commercial machines of a fairly wide variety.

The Jacobs Windmill

From 1930 to the middle '50s it was possible to buy American-made electric windmills. The outstanding model was built by the Jacobs Wind Electric Company.

Jacobs began experimenting with windmills in the early '20s and by 1930 had developed a 15-foot diameter, 3-blade rotor with matching generator of 2500 watts at 32 volts or 3000 watts at 110 volts. The generator weighed 440 pounds with a 9-inch diameter armature and a 9-inch core length. There were 60 pounds of copper wire in the field coils. The generator was direct coupled to the rotor shaft. It was designed to operate at a speed range of 125 to 225 rpm, or in wind speeds from 7 to 20 mph.

In a locality where there were wind speeds of 10 to 20 mph for 2 or 3 days a week the output of the plant was 400 to 500 Kwhr per month.

For charging the batteries a special step controller was developed. This controller enabled deeper charges to be put into the batteries than any other known controller.

From records kept on more than a thousand plants over a 10-year period it was found that repair costs averaged only $5 per year per plant. The operating and maintenance costs were largely limited to the replacement of batteries, which, over a 10-year period, amounted to about $36 a year.

In 1975 dollars the installed cost of the Jacobs plant, with batteries, would be about $800 per Kw.

The Jacobs design concepts

An interview with M. L. Jacobs, American windmill pioneer, revealed a number of ideas often overlooked by less experienced designers.

The noise problem is a case in point. Often overlooked by windmill advocates is the fact that these machines are potentially noisy. Any object moving rapidly through the air has a tendency to generate vibrations in the audible range, and rotating blades are no exception. If windmills were always located in remote areas, noise would be no problem; but, if wind is to be a significant energy resource, windmills and people must do a lot of co-existing.

Windmill blade designers turn to aircraft airfoil designers for help because of the similarity of functions, but the aircraft designers can offer no analysis techniques for the elimination of sound. It is likely that noise vibrations can be incorporated into the windmill blade equations if the effort is made.

The Jacobs blade has a flat face and a curved back giving a profile with the traditional thick leading edge and streamlined contour. While the contour of the Jacobs blade was empirically derived, the objective of attaining a silent rotor was attained.

On the question of where the rotor should be placed, Jacobs presents some compelling arguments for placing it upwind. Those who favor putting it downwind claim that the wash of the rotor blades, retreating from the upwind position, puts a large burden on the tower. Jacobs says that a properly designed rotor presents essentially a solid front to the wind. When a blade of such a rotor is turned to the proper angle of attack, there is a certain volume of air contained in the cube consisting of the swept angle of the blade and the thickness of the angle of attack. The volume swept by a blade during the time it takes for the next blade to arrive at the same position is the same as the volume pushed into the blade by the wind. As a result, the rotor presents virtually a solid front to the wind, and the maximum power is extracted.

Since power has been removed from the wind, its speed is reduced. The air flow back of the rotor—that which strikes the tower—is a gentle breeze compared to what would strike it if the rotor were placed downwind.

Jacobs also points out that the tower produces a turbulence that seriously affects the rotor when it is placed downwind. Each time a blade passes through the wind shadow the turbulence effect produces a vigorous longitudinal vibration on the down blade in contrast to the up blade, which is above the tower. As a result, severe bending and vibrating stresses are placed on the shaft and bearings, the fatigue problems are worsened, the lug bolts holding the blades to the hub suffer embrittlement, and in some cases the generator windings are even torn loose.

Considering the effect that turbulence can have on a windmill, one may wonder whether turbulence is a limiting factor in maximum size. Turbulence is generated at ground level. The wind flow at higher levels is fairly smooth. In a large windmill the turbulence difference between the up blade and the down blade would be quite substantial, and the resulting vibratory stresses quite great. If the rotor is small enough so that most of it is subjected to the same set of forces, that source of vibration is reduced. The turbulence problem for large rotors could be reduced by placing them on very high towers, but the cost of higher towers would rapidly put windmills out of cost competition.

Wind tunnels are useful instruments for the design of airfoils. But wind tunnels produce laminar flow, whereas nature produces eddies and gusts of completely random direction and intensity mixed with laminar flow. Unless wind tunnels having a greater degree of sophistication are invented, the ultimate test for power and noise and vibration will still be in nature.

The Jacobs blade is constructed of airplane spruce and painted with a special aluminized paint. This surface resists the accumulation of ice and snow.

The Jacobs blade has a twist to match the thrust to the differing radial velocities along the length of the blade.

Since with properly designed blades the rotor presents an equivalent solid surface to the wind, there is no work advantage to using more than two blades. Greater power output is not attained by increasing the number of blades. The Jacobs windmill uses 3 blades, but the reason is for balance, not for power. The two-bladed rotor was subjected to excessive vibration when it turned to head into a new wind direction.

Analysis of the vibration problem is as follows: When the 2-blade propellor is in the vertical position it offers no centrifugal resistance to the tail vane, but, spinning into the horizontal position, it offers 1100 pounds of gyroscopic resistance to the tail vane movement. With the 3-blade propellor when one blade is in horizontal position two blades are above or below horizontal by 60 degrees thereby reducing the gyroscopic resistance of the one horizontal blade. The resistance to the tail vane control is smoothed out, and the vibratory oscillations disappear.

The windmill output power rises on a slow curve with increasing speed, whereas the power curve for conventional electric generators rises on a fast curve. Jacobs points out that most windmill designers neglect the matching requirement. His windmill generators are designed to match the rotor power curve and as a result are much heavier than most generators.

Generators mounted on the top of high steel towers are subjected to heavy electrical discharge from lightning pick-up effects and from static build-up on the blades. It is necessary to avoid the static discharge into the direct coupled armature and to prevent nearby lightning discharges from arcing between bearing and races and causing weld spots.

Jacobs found that the ways to correct these problems were to use extra heavy insulation, to couple a large capacitor across the generator brushes and frame and to add dual sets of heavy grounding brushes on the armature shaft. With these features he never had to replace a generator in 20 years from lightning damage or other burn-out.

The Jacobs Wind Electric Company stopped making windmills in 1957 because the rural electrification program greatly reduced the demand. Jacobs himself keeps semi-active by doing consulting work in the field. However, with the greatly reawakened interest in windmills, he expects the company to be reactivated and to offer a windmill in the 10 Kw range.

The Princeton Sailwing

With the coming of the energy shortage and certain other social and economic squeezes there has been a burst of inventive energy applied to windmills. An adequate survey of all of the interesting ideas would require a book in itself.

The Princeton sailwing rotor is one of the more practical of the new ideas. The Flight Concepts Laboratory of Princeton University became interested in windmills as a result of research performed over the years on the sailwing airfoil. This device was first conceived as an advanced sail for a boat and later was applied to a wing for an aircraft. The sailwing has the aerodynamic characteristics of a well-designed rigid aircraft wing, and it has actually been flown on low-speed aircraft.

The first Princeton windmill with sailwing blades was constructed in 1966. It was a 2-blade, axial flow type 10 feet in diameter. This windmill was originally intended as a research device for the study of the sailwing aerodynamic characteristics at various blade pitch and twist angles under all wind and weather conditions over the four seasons. This machine was tested over a period of a year, withstanding without damage many gale force winds, freezing rain and heavy snow. The structure, including the Dacron sails, appeared at the end of a year to have suffered no significant wear or ill effects.

The blade of the sailwing is constructed with a rigid leading edge, tip and root section. The tip and root are connected by a trailing edge cable fastened to a wrapped around (two-surface) sail. Unloaded the sailwing has no significant contour, but when loaded it takes on the contour of a well-designed airfoil. The construction is illustrated in Figure 11-1. The sailwing offers lighter weight and lower cost than almost any other type of construction.

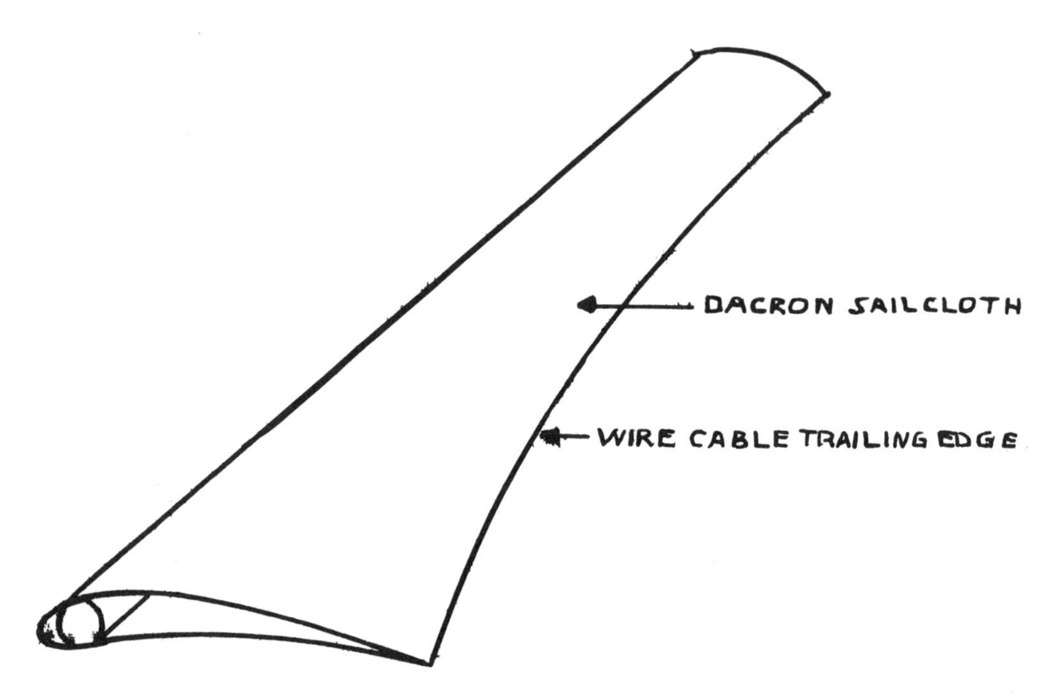

Figure 11-1. Princeton sailwing design features.

In 1972 Princeton built and tested a 25-foot model of the sailwing windmill. The light weight of the blades eliminated the gyroscopic effects usually expected from a mill of this size. The power output of the 25-foot model is 6 Kw at 20 mph winds and 20 Kw at 30 mph winds. T. E. Sweeney of Princeton estimates that 50-foot sailwing windmills could be built without trouble, but he is unprepared to extrapolate farther.

The Princeton sailwing windmill is being investigated by Grumman Aircraft for possible commercial development.

The Darrieus Vertical Axis Rotor

Most of the windmills invented before the Dutch horizontal axis windmills of the 12th century were of the vertical axis type. It is likely that windpower never became very important in the pre-industrial world because the tip speed velocity ratio of these devices was always less than the wind speed. As a result, the efficiency was poor and the power output was low. With the invention of the modern lift-producing blades, the horizontal axis windmill, achieving velocity ratios of 6 to 7, have become very efficient.

The vertical axis windmill continues to hold interest because the horizontal type requires complex and expensive drive, yaw and rotor control mechanisms. The vertical axis mill responds to wind from any direction and instantly to wind changes without any control mechanism. Further, most horizontal axis designs require that the power be taken off the center axis, presenting difficult installation and maintenance problems; the gondola containing the drive shaft, gear box, generator and control mechanism must be mounted at the top of the tower, perhaps 100 feet or more in the air. With the vertical axis rotor, power can be taken off at ground level.

In 1931 G. J. M. Darrieus patented a device looking like almost anything except a windmill. If one were to pierce a barrel hoop with a shaft at the two diameter points and rotate the hoop around the shaft, one would have the appearance of a Darrieus windmill rotor. See Figure 11-2. The difference between a hoop and a Darrieus rotor is that the latter is a well-designed airfoil.

The reason for the circular shape can be understood by imagining a square instead of a hoop rotating on a shaft. In the insert of Figure 11-2 the sides of the square are airfoils, and the cords to the shaft are simply brace members. This device will rotate in the wind. However, at any significant wind speed the centrifugal force will cause the vertical airfoil members to bend outward unless heavily braced.

There is a very simple solution to that problem. If the vertical members were limp strands, rotation would throw them into a catenary subject only to tensile stress. The rational thing to do is to design the blade into the shape of the catenary to begin with. Hence the peculiar shape of the Darrieus rotor. Sometimes the machine is fitted with three blades instead of two to increase the solidity.

The Darrieus rotor has one fault. It is not self-starting.

At present there are at least four places experimenting with the Darrieus rotor. One is the Low Speed Aerodynamics Laboratory, National Research Council, Ottawa, Canada. NRC constructed a 14-foot model and tested it in a wind tunnel at wind speed of 12 mph, varying the solidity with 2-blade and

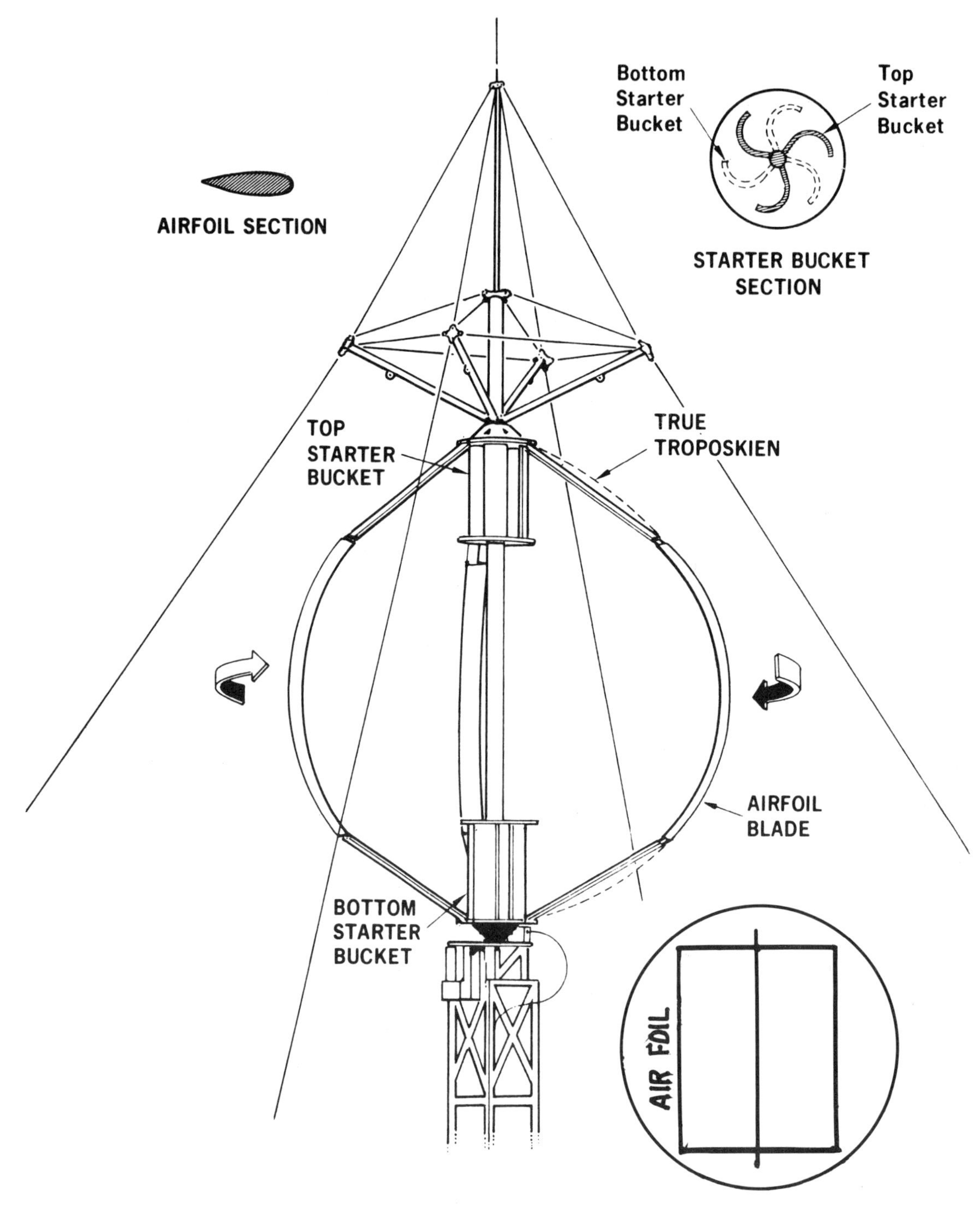

Figure 11-2. The Darrieus Rotor.

3-blade configurations. The maximum output of the 2-blade model was 0.67 h.p. at 150 rpm, and of the 3-blade model it was 0.65 h.p. The rotation of the 2-blade model was 150 rpm and the 3-blade model was 130 rpm at the 12 mph wind speed. The velocity ratio, or tip speed, was 6, and the efficiency was 70%. In estimating power from the Darrieus rotor, the swept area must be reduced by 1/3 because the ends of the blades contribute little power.

NRC has since completed two 12-foot models and has installed them on the roof of the laboratory. One important change has been made. Instead of the continuous curve, the blades are made in three straight sections approximating the catenary curve. This change makes for easier manufacture and transportation.

NRC has purchased a number of preproduction 15-foot units from Dominion Aluminum Fabricating Limited (3570 Hawkestone Road, Mississauga, Ontario, L5C2V8). This company is manufacturing Darrieus rotors and will be selling production models by mid-1975.

Sandia Laboratories (Albuquerque, New Mexico 87115) seems to be the most active group working on the Darrieus rotor in this country. They have applied for patents on improvements to this mill.

One of the patents applies to a curved member in the mid-section with straight sections for each end of each blade. This approach compares with the Canadian version, which uses straight members for all sections.

Another Sandia improvement is the placement of a starter bucket on each end of the central shaft to start the blades rotating automatically when the wind blows. Each bucket consists of three C-shaped fins oriented in different vertical planes for smooth starting.

Langley Research Center (Hampton, Va. 23365) is doing some research on the Darrieus rotor.

Research and Design Institute (P. O. Box 307, Providence, R. I. 02901) specializes in the "re-cycling" of old buildings on the thesis that a building represents stored energy that should not be wasted. They are restoring an old mill to use as their headquarters. One of the activities at their new office will be to experiment with sun and wind energy to determine whether a building can become self-sufficient in its power needs. They are building a Darrieus rotor windmill as a part of their experimental program. It is interesting to note that they are using a Savonius rotor at the top and bottom ends of the vertical shaft for self-starting.

Savonius Rotor

This version of the vertical axis rotor is the best known. The construction is essentially that of an S-shaped airfoil with a shaft down the center of the S. One can imagine the shape of a Savonius rotor by thinking of an oil drum cut lengthwise in half with one half turned around to create the letter S. In fact, crude Savonius rotors using split oil drums have been successfully built. This rotor was invented by S. J. Savonius in 1929. It has found successful application as an ocean current meter, where it can accurately measure ocean currents speed as low as 0.5 knot.

The Savonius rotor is self-starting in wind from any direction, but its velocity ratio is low—0.8 to 1.8. Due to the poor velocity ratio the Savonius rotor has received little research attention. As a result, the development alternatives have not been intensively examined, and little is known about the maximum potential design. Some of the possible alternatives are:

—Variations in height
—Variations in aspect ratio, or ratio of height to diameter
—Variation in number of blades, 2, 3 or 4
—Variation in blade profile. Should the blade profile be such that the diameter is greater at the bottom to accommodate the lower and slower winds?
—Variation in chamber depth. Should the cup of the S be a full half-circle or a more shallow configuration?
—Variation in blade thickness
—Variation in blade contour. Should it have an airfoil design?
—Variations in the arrangement of the chambers. The usual arrangement is to attach the ends of the chamber to a shaft. However, the shaft may be omitted by using the end plates to hold the chambers in place. In this case, the chambers may be separated and overlapped. How much separation and overlap is desirable?

One question seems to have been missed by Savonius rotor investigators: What is the overturning moment of a very tall rotor? Could it be contained by guy wires?

The Chalk Ground-Mounted Power Take-off

For a number of practical reasons the power conversion generator of a horizontal axis windmill is mounted at the top of the tower where it can be connected to the rotor drive shaft. As a result, the tower of a large windmill must be designed to carry the overturning moment of several thousand pounds in addition to the drag load of the rotor. To obtain less stringent tower requirements, less expensive installation and easier maintenance, it would be nice to have the converter at ground level.

One new horizontal axis design idea can have the power take-off at ground level. It has attracted attention for its resemblance to a bicycle wheel and for its possible low cost. This design arranges the blades around the wheel like spokes. This design has both a hub and a rim. It is quite easy, then, to couple the generator to the rim at the ground level.

The Chalk bicycle wheel windmill at first seemed to offer a cost breakthrough. Using a rim for bracing, it was possible to construct the blades of thin aluminum. But, if this windmill is a good idea it still needs much engineering. For example, it should not be steered by a tail vane; its heading should be controlled by a servo mechanism. This windmill suffers from the complaint of all multi-blade machines: increasing the number of blades does not increase power output because increasing the solidity ratio beyond 0.2 to 0.3 causes power reduction.

The Chalk windmill introduces the possibility of considering the rimmed rotor, an idea that seems not to have been examined before. Retaining the rim but reducing the rotor to three airfoil blades seems worth examining. The rim would introduce rotor stability problems in a large windmill, but it seems worth a study to see whether the trade-off between these problems and the advantage of low-level power take-off would be a gain.

The Princeton Auto-Rotating Vane Windmill

It has been known for many years that a flat plate will, when pivoted about its mid-chord axis, auto-rotate when subjected to sufficient wind velocity. Such a flat plate, however, will not always be self-starting, and it will operate in either direction. To overcome these objections it is only necessary to add vanes to either side as shown in the plan view of Figure 11-3.

The Princeton experimental version of this idea is a long, narrow vane 5 feet high and 1 foot wide mounted on a vertical shaft with an automotive alternator at the bottom end of the shaft. Princeton has not been experimenting with this device long enough to define any of its characteristics or do any analytical calculations. They have found it capable of producing useable power.

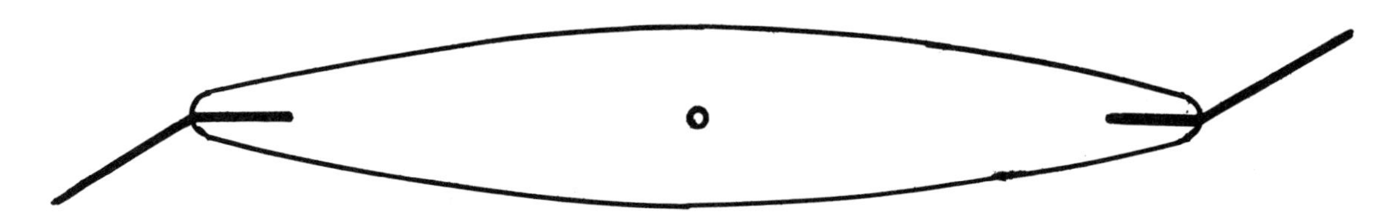

Figure 11-3. The Princeton auto-rotating vane.

The auto-vane windmill may turn out to have the same weakness as the Savonius rotor, a slow velocity ratio. However, a device of this sort ought to be very inexpensive to build. It might be possible to gang them up in arrays to produce a sizeable amount of power at a cost per Kw comparable to the more efficient horizontal axis type.

Shrouds and Diffusers

Placing the rotor within a duct can offer advantages under some conditions. The preferred design is a shroud over the rotor having a diameter only slightly larger than the rotor and a diffuser section which gradually expands back of the rotor. The function of the diffuser is to convert the kinetic energy of the flow downstream of the rotor into a pressure rise, which lowers the pressure level behind the rotor. The rotor is then able to capture air from a free stream area that is larger than the rotor itself. The inlet area need not be large if the diffuser is sufficiently effective. The flow velocity through the rotor is 20 to 60 per cent greater than the free wind velocity.

The diffuser-augmented wind turbine gives up to twice the power of an unshrouded rotor according to some investigators. There are advantages at both ends of the speed range. Since the air flow is speeded up, more power can be extracted than usual from low speed winds. At high speeds the diffuser offers the opportunity to accommodate without the use of variable pitch blades.

Grumman Aerospace Corporation (Bethpage, New York) has done some investigations on the diffuser-augmented turbine.

THE TRACKED VEHICLE SYSTEM

Probably the oldest application of wind energy is to the sail boat. The sail boat idea can be extended to land. Just put airfoils on a vehicle and the vehicle on a track, and the breeze will send it merrily along. The sail is not a very practical idea for a railroad, but now comes the idea of attaching an electric generator to the axle of the vehicle. To complete the idea, the track is made into a loop and the loop is entirely filled with carriages so that the whole system of cars moves as a unit.

Montana State University and Mississippi State University have done some research work on this idea. They have done a computer analysis of the necessary design characteristics and have arrived at some estimates of the power output. Their calculations show that a track 8 km long with airfoils 12 m high could supply the electrical needs of 15,000 people. Their findings also suggest that the maximum wind energy extraction rate may be greater than conventional windmills by 20%.

A practical demonstration of these research results has not yet been made. The Public Service Commission, Burlington, N. J., conducted experiments on this type of scheme in 1932, but data on the experiments were not available to this Report.

One of the objections to the tracked-vehicle idea is the amount of open land required to produce any significant amount of power.

Dr. Pasquale M. Sforza of Polytechnic Institute of New York (Route 110, Farmingdale, N.Y. 11735) has under development a "Vortex Augmentor" that will double the input wind speed.

12

UTILITY EXPLOITATION OF THE WIND ENERGY RESOURCE

THE KILOWATT YEAR

The yearly output of a combination of specific site and specific windmill is predictable with a great deal of accuracy and reliability. For example, if there is a site whose wind duration curve shows an 18 mph wind for at least 50% of the year, a windmill with a 100 Kw rating at 18 mph will produce 438,000 Kwhr during the year.

This concept of the yearly yield suggests the usefulness of a new unit of measure for windmills, the kilowatt year. Let this unit be defined as the number of kilowatt hours generated per year per kilowatt of maximum design capacity.

Most wind machines are today specified at a maximum rated output. There is a cut-in speed where some power is generated. The power increases up to a maximum design speed and then remains constant by means of rotor control. The velocity duration curve may show a maximum rated power wind speed of, say, 18 mph for 50% of the year. It might also show wind speeds of 25 mph for 20% of the year, but all the power in the wind between 18 and 25 mph will be allowed to escape.

The wind power increase is linear from about 25% to full power. If the speed duration curve shows that the rotor could produce 25% of the rated power for 10% of the year, 50% for 10% of the year, 75% for 10% of the year, and 100% for 50% of the year, the yearly yield per installed Kw is, respectively, 219 Kwhr, 438 Kwhr, 657 Kwhr and 4380 Kwhr. The yearly harvest is 5694 per installed Kw.

From these numbers it can be shown that for this example the windmill produces about 25% of its yearly power at speeds below rated power, or 75% of its yearly yield at the rated power. Most cases will show a close parallel to this typical case. The kilowatt year per kilowatt capacity is 4380 Kwhr. Since most of the derived energy comes during the time when the wind speed is delivering maximum rated power, the kilowatt year may be normalized to this value. Assuming the ideal condition where the site delivers rated wind 50% of the time, the value of 1 Kwyr is 4380 Kwhr.

The Kwyr becomes a good yardstick for various measurements. For example, if the windmill costs $1000 per Kw to install, the cost is $4.38 per Kwyr per kilowatt of capacity.

This measure may also be used to judge the best size machine to buy. Some have argued that it is more practical to install, say, ten 10 Kw mills than it is to install one 100 Kw machine. Assume that the 100 Kw machine is rated at 18 mph and the 10 Kw machine at 15 mph. The velocity duration curve shows winds above 18 mph 50% of the time and above 15 mph 60% of the time. The yield of the large machine is 1 Kwyr per Kw because it produces full output 50% of the time. The smaller machine, because it operates at full capacity for 60% of the time, yields 1.2 times as much energy. However, the smaller machine may cost more to attain equivalent power rating because the rotor has to be larger to attain the same power at lower wind speeds. If the large machine costs $800 per Kw and the small machine costs $1000 per Kw, the buyer would be paying 1.25 times as much for the smaller machine to gain 1.2 times as much energy.

Combining Site and Windmill

Examination of a number of speed duration curves shows that the mean annual speed occurs about 50% of the time. A case can be made for installing windmills whose rated output is at the mean annual speed. However, the maximum output of a windmill is attained as a trade-off among a number of design factors, such as rotor diameter, blade twist, tip speed, solidity ratio, number of blades, generator capacity. It is not always possible to optimize all of these factors at a preferred design speed.

The fault with a mill having a maximum power at the mean annual speed is that the power above the mean is lost. There is 1000 times as much energy in the wind at 25 mph as at 15, and 340 times as much as at 18 mph. However, at a site with an annual mean of 15 mph the 25 mph rate probably does not occur more than 10% of the time. Although the annual yield even at the short duration would be 100 times more than would be obtained at the design speed of 15 mph, the need for continuous delivery of power at a fixed rate overshadows the advantage of designing for higher speeds.

The trade-off preferred by most designers places the maximum power point at some location higher than mean annual speed but not so high as to reduce the mean annual power. This compromise seems to be at about the speed where the wind blows for about 35% of the time. In most regions the annual mean is about 15 mph and the 35% mark is at about 18 to 20 mph. Hence, the popular design speed is 18 to 20 mph.

An acceptable windmill site would, from these considerations, be one where the maximum power speed was sustained for 35% of the time, or where the mill rating was 0.7 Kwyr per Kw.

Cost versus Size

Windmill designers strive for larger and larger rotors because the power increases with the square of the diameter. Since the designer can not do anything about wind speed, which increases the power with the cube, he seeks to make up the difference by increasing rotor diameter. If the cost of the mill goes up linearly with the size while the power output goes up with the square, the large mill designer has one up on the small designer; he can increase power faster than he increases cost.

Whether installed costs do vary linearly with size is not answered accurately at present. It seems likely that machines of the 100 Kw size or larger will never be built in mass production. The design may become standardized, making it difficult for the buyer to tailor his design to the site. But it is likely that the large machines will be built in batch quantities. If the buyer contracts to buy an entire lot, he probably will be able to tailor some of the design factors to his sites. This service will cost money. Even if the large machines are built to standard specifications, they will be manufactured in batches, which is more expensive than mass production.

On the other hand, windmills of the 10 Kw size are likely to be candidates for mass production. It is known that the householder wishing to be independent in all of his electrical needs should have a capacity of about 10 Kw. Could the mass production of 10 Kw sizes result in an installed price per Kw equal to or less than the larger sizes? No one seems to know for sure at present. Experience is lacking.

Cost versus Weight

South and Rangi of the Canadian National Research Council have used an interesting method of cost comparison. By plotting the weight of a number of different machines in various sizes they found that the weight per square foot of swept area of horizontal axis windmills is always in the order of 12 pounds per square foot. The cost of these windmills has varied from 80¢ to $1.25 per pound. At the higher price these windmills are about $15 a pound per square foot of swept area. Within a narrow range of individual variations, the cost of horizontal windmills does vary linearly with size, according to these investigators. It must be remembered, however, that there are no up-to-date figures available comparing batch production and mass production of these items.

South and Rangi use this method of estimating to justify the value of their catenary design. Their estimated weight of the catenary windmill is in the order of 1 pound per square foot of swept area up to 3000 square feet, with the weight increasing from that point at a rate proportional to the square root of the swept area. If the weight estimate were raised to 2 pounds per square foot for safe estimating, at the cost of $1 per pound per square foot, the catenary windmill will cost about 1/6 as much as the horizontal type.

The development of this type of windmill will bear watching, particularly in the smaller sizes.

Windmills Coupled to the Utility Grid

The utility engineer may wonder how an intermittent and uncontrolled power source such as a windmill can be practical.

Most current thought regards storage as a necessity for intermittent wind energy. For the utilities no such storage is needed. The windmill is simply tied into the grid, and it supplies power as long as the wind blows.

Since the power unit is small compared to the total load, it should be connected into the low voltage distribution system wherever possible. This practice may limit the siting of windmills in some cases. Tying into the low voltage system avoids substation construction and high transmission costs.

Experience so far has indicated that it costs more to produce electricity with wind than by conventional means. The ease with which the utility can distribute wind power offsets the higher costs to some extent. If the wind energy contribution does in fact increase the total cost per Kwhr in a system, there is already in effect a method for handling the situation. The "fuel charge" added to the electric bill is a device for covering expenses over standard costs. The difference due to wind costs could be a part of a similar fuel charge. The wind charge would no doubt be less than the fuel charge, and it would not vary with the market price of fuel.

Windmill control on the utility grid would be easy to handle. Yaw control, pitch control of the blade, generator cut-in and switching to the line would be automatic. They would be a part of the windmill installation and would be handled by windmill maintenance. The windmill would not be tied into the central computer control. The windmill would simply feed into the grid whenever the wind blows. The central computer would need to know when and at what rate the wind plants were feeding the line, but this requirement would be met by some simple information transmission system.

The Wind Energy Harvest

How much wind energy can a utility harvest? Consider the case of a region with an installed capacity of 5 million Kw and a delivery of 25 billion Kwhr per year. Assume that this region has 100 acceptable sites. Recall that an acceptable site is one delivering 0.7 Kwyr in 35% of the time.

It is not yet known how many windmills can be installed at a site, but for estimating purposes assume it is two. Assume that the machines installed are of the 100 Kw size. The capacity of the 100 sites is 20,000 Kw, or 0.4% of the region's installed capacity.

Since the rating factor of these windmills is 0.7 Kwyr per Kw, the 20,000 Kw capacity yields 14,000 Kwyr. Converting to Kwhr (multiplying 14,000 by 4380), the yield is 61.32 million Kwhr per year, or 0.25% of the total delivery.

WHAT SHOULD THE UTILITIES DO ABOUT WIND POWER?

The numbers derived above do not suggest a great promise in wind power to the utilities. If the 1 megawatt machine is developed, the capacity and yield percentages would be increased by 10. However, there is no certainty that the 1 megawatt machine will be successful.

Before the utility manager discards wind completely he should be sure he has explored every angle.

Are there sites where the mean annual wind is perhaps 5 mph higher than most places?

If there are a few very good sites, what are the chances of putting up a cluster of several mills?

Is there any chance of contouring the land or building diffusers so as to increase the normal wind speed?

Would it be possible to use windmills to electrolyze water into hydrogen to burn in fuel cells?

In the coming years no energy resource is too small to be ignored.

THE MODEL COMMUNITY AND WIND POWER

The model community discussed above used 86 million Kwhr per year of electricity. By installing solar cells on roofs, side walls, parking lots and back yards, a surplus of electricity could be generated. It is obvious, of course, that not all of the available areas could or would be devoted to solar cells. A deficit instead of a surplus could easily prevail. It would be a good idea to use wind as a complement to solar power.

One advantage of wind power over solar power is that the wind blows any time, often at night.

Assume that this community has one major acceptable wind power site. If two plants were installed at this site at a rating of 0.7 Kwyr per Kw, and if the plants were of the 100 Kw size, the potential yield is 613,000 Kwhr, or 0.7% of the community needs.

Looking for other windmill locations, consider the residences and business buildings. Perhaps each residence could handle a 5 Kw mill and each business a 15 Kw mill.

The trouble with this idea is that only a few residences and buildings are suitable for windmill installation. The neighbors are too close, there are too many trees, and complications involving codes and zoning limit the number of acceptable locations.

Assume that 1000 residences could be found for 5 Kw mills and 25 business buildings for the 15 Kw size. Assuming the machines to have a 0.7 rating, the total yield for these machines per year is 16.5 million Kwhr, or 19% of the community needs. Combined with the 100 Kw machines, wind power could supply 19.7% of the community's electrical needs.

THE COONLEY WIND BUILDING

Douglas Coonley, in his M.I.T. Thesis (Design with Wind, 1972) has offered an interesting idea relative to the use of buildings to collect wind power.

Tall buildings divert the mainstream flow of the wind and cause complicated air flow patterns. A large downwash vortex occurs at the base of the building on the windward side. Large, lazy eddies form on the leeward side. Rising vortices form around sharp corners. Suction from rising vortices draws air from all directions, resulting in strong winds near the ground. In addition to the stress problems on structural members, wind forces cause difficulty in operating entrance doors, render ineffective the protective awnings, produce adverse effects on air conditioning and ventilating intakes, and increase the hazards to window cleaners and maintenance staff.

Coonley found from his wind tunnel tests that openings or "slots" through buildings greatly reduced the wind problems. One can think of a Coonley slot by imagining the top two stories of a building being placed on pedestals above the rest. Elevators, support members, wires and service tubing go through the pedestals. The wind blows through these open spaces instead of being thrown into vortices. The building becomes a permeable structure instead of an obstacle. The wind velocity through the slot increases 2 to 4 times over normal speed, a fact that probably accounts for the reduced pressure on the rest of the structure.

It then occurred to Coonley to put windmills in the slots of the buildings. Windmill installations on the top of a new building could be done, but the disadvantages would be: need of a tower to get above the wind shelter area and edge diversions, no relief of wind pressure on the building if the slots were omitted, and greatly increased stress on the building from windmill drag. Place the windmill in the slot and gain these advantages: lower stress on structural members, no tower required, and two-point support for the rotor.

Let it now be assumed that a 30-story office building, size 100 × 300 feet, is to be built in the model community. At an estimated 10 watts per square foot, the electrical needs of the building would be about 4 to 5 million Kwhr per year.

Assume that two stories of the building were opened to accommodate a windmill slot. This space could utilize windmills of about the 15 Kw size. If the prevailing wind had a mean speed of only 10 mph, at the height of about 300 feet where the mills are mounted the speed would be at least 15 mph. If, as Coonley found, the speed increases at least two times when going through the slot, the speed at the rotors would be at least 30 mph. The design point for this size windmill could easily be 30 Kw. As many as eight of these machines might be installed. The potential yield of the windmills is then 4.2 million Kwhr per year.

Coonley has not examined the added cost to a building to provide a slotted support for windmills. Assuming the building residents were paying 3.5¢ per Kwhr for electricity, the value of the windmill power would be $2.94 million over 20 years. If the cost of the windmills were $1000 per Kw, the total would be $120,000 (using the nominal 15 Kw rating for the calculation). Thus, $2.82 million would be left for construction of the slot.

13

ODDS AND ENDS

SECTION 1. ENERGY FROM OCEAN THERMAL DIFFERENCES

There Is Power in Ocean Heat

In tropical and semi-tropical areas there is a temperature difference of 20 degrees C between surface water and water at great depths. The theoretical Carnot efficiency of this temperature difference is 6% to 7%, and the achievable conversion is probably one-third to one-half of that amount. However, if a large volume of water is circulated, significant amounts of power can be harvested.

In the open cycle method used by Claude in the 1920's, the warm water was flash evaporated in a partial vacuum, sent through a turbine and the vapor condensed by the cold water from the depths. The preferred method today is to use ammonia or propane in a closed cycle.

Great quantities of power from this source are predicted by its advocates. Anderson and Anderson, for example, suggest that 182 trillion Kwhr per year could be extracted from the thermal gradients of the Gulf Stream alone. The costs, too, are very competitive. One estimate places the cost of a 100 Mw plant at $16 million, or $166 per Kw, and the cost per Kwhr, including maintenance, at 8 mills per Kwhr. These plants will probably be modular, consisting of 25 Mw units. They will be clustered to obtain any capacity desired.

This idea is scientifically sound and has been proved by both experience and study. Between 1947 and 1953 the French had a successful 1500 Kw plant operating off the Gold Coast. No significant technological breakthrough is required. Most of the remaining problems are of an engineering nature. Many of those problems once considered formidable now have feasible solutions. For example, deep hole drilling in water depths as great as 6000 feet has showed that long pipes can be suspended in deep water; undersea technology is well enough advanced to show that underwater construction, operation and maintenance is possible; stable floating platforms have been built; undersea transmission of electricity is practical and is in use in various parts of the world.

One of the technical problems is marine fouling on the warm water heat exchanger. A study at the University of Massachusetts has shown that a current of 2.2 knots across a surface inhibits marine growth. Thus, it is possible to use natural forces to overcome one of the inherent problems.

Woods Hole Oceanographic Institution has found that a very small concentration of chlorine prevents microbial growth. When chlorine gas is added to the sea water it forms a hypochlorous acid, which can also be formed directly by the electrolysis of sea water. The power required for this operation would be very small. Such equipment has already been developed for the intake pipes of power plants operated on the sea coast.

Concern has been expressed for some legal and social issues, but this anxiety may not prove valid. For example, the precedents for anchoring anywhere and for laying cables or pipes in the sea bed without contention are well established. These systems, if installed off the Florida coast, as presently planned, will not interfere with established commercial and sports fisheries.

Would raising this much cold water from the depths change the surface temperature? Those who have studied this question have not been able to justify this fear even if huge amounts of power were withdrawn. If the world population of 2000 were supplied electricity from ocean resources at the present U.S. per capita rate, the tropical surface temperature would be lowered less than 1 degree C. It is known that natural causes have created temperature changes of 1 degree per year in large bodies of water without destroying ecological balance. That amount of change, however, should probably be considered the upper allowable limit of change induced by this system.

The question that seems not to have been answered yet is whether these stations are to be manned or unmanned.

National Science Foundation has been giving increasing attention to this energy resource. Beginning with a research funding of $84,100 in 1972, it has increased its funding to $3 million in 1975. Carnegie-Mellon University and University of Massachusetts are continuing their theoretical studies, and several industrial firms have contracts dealing with practical problems, such as the design of heat exchangers. University of Hawaii has a contract to study the potential for a plant in Hawaii.

It is likely that a pilot plant can be built by 1980 and a large scale plant by 1985. It is likely that several hundred Kw capacity per year could be added after 1985 until all of the practical sites are used up. The number of possible sites has not yet been determined. Florida could be supplying one-fourth of its electrical needs from this source by 2000.

SECTION 2. BIOCONVERSION OF SOLAR ENERGY

Everyone Eats Solar Energy

Every living organism converts and stores solar energy in some form during its metabolic processes. These organisms can be made to give up their stored energy.

The most common bioconversion recovery process in use today is the pyrolysis system. This process burns biological material in an oxygen-starved chamber. The product is either methane or a synthetic oil. The process produces about four times as much energy as it used in combustion. Several companies offer different but basically similar pyrolysis hardware.

There is some possibility of bioconversion directly to electricity or hydrogen, but these ideas are only in the investigation stage.

The main sources of materials for bioconversion are certain rapid growing marine and land crops, agricultural and logging residue, animal feed lot waste and urban waste. The extent to which energy farming will be practiced for the deliberate growth of bioconversion materials is not known. The basic efficiency of bio-solar conversion is only 1% to 2%.

The need for human food may severely limit the amount of land that can be allotted to energy farming. Agricultural and logging wastes can be made into a variety of commercial products. On the basis that it is wiser to transform any material into a new material than to convert it to energy, agricultural and logging waste should not be used as feed stock for pyrolysis.

However, not all of these wastes can be used to produce new materials; conversion to energy in many cases is better than letting the stuff lie to rot. Animal feedlot wastes are a source of serious pollution, since they are no longer used to any extent as fertilizer. Conversion to methane gas is a desirable goal.

However, a process is being developed to convert these wastes into a protein that can be used as animal feed. If successful, this use should have precedence over energy.

Urban waste disposal has to compete with recycling interests. A case can be made for pyrolysis as a method of recycling when combined with proper sorting techniques. Modern plants recover metals and glass for recycling and pyrolyze the rest.

One of the important bio-sources of energy is animal fat. The significance of this hydrocarbon seems not to have been studied. How much food energy ends up in animal fat is not known, but great amounts of it end up in the garbage, and the best way to recover it is by pyrolysis of garbage.

The utilities could profit from bioconversion because it would reduce their dependence on fossil oil. It is not likely that utilities can do much about the pyrolysis of urban waste, however. Waste disposal is a problem for public action and is heavily laden with controversy and politics. Pyrolysis, though its product is cost competitive, is capital intensive. Since the capital outlay comes out of already overburdened taxes, it is very difficult to move toward this resource.

National Science Foundation is spending $5 million in research in this area. There are at least three large firms offering proven pyrolysis systems, and there are a smattering of communities starting to use this method of waste disposal.

Bioconversion resources could supply about 2.5% of the national energy needs by 1985 if development were systematically pushed.

SECTION 3. ENERGY FROM THE PROPERTIES OF MATTER

The thermo-electric effect, where electric current flows between the hot and cold junctions of dissimilar metals, is a property of matter, but as a power source it has not attained significance.

The property of potential difference between metals is widely used in batteries. The direct conversion "dry" battery is used by the millions in small apparatus, but it has no significance to the utilities. The storage battery will probably become important to the utilities within the next 10 years, but it is not examined in this Report because it is considered a secondary, not a primary, energy resource.

Nuclear Power

Nuclear power is not examined in this Report because utility management can obtain information about nuclear fission from official sources without difficulty. Some comments are in order, however, about the relation between nuclear power and other resources.

Conventional fission has usually been offered as the best hope of counteracting the fossil fuel pinch. It is well known that fission plants have slipped badly in installation schedules. It is also well known that it takes ten years to build a fission plant. Fission plants planned today will not be on line before the middle of the next decade. It is suspected that uranium ores for conventional fission will be exhausted in 50 years at the present rate of consumption. If a massive effort were launched to build a large number of these reactors, the chances are that the fuel would run out before they had expended their useful life.

The breeder reactor has been promised as the great hope until fusion is attained, but it may turn out to be a sad waste of research money. Billions have been spent on it, and it is not much closer than fusion. It may be less acceptable socially than conventional fission. Even Edward Teller himself has pointed out that in the end the breeder reactor may be just too expensive to build.

Fusion is much closer than most people think. In a paper read at the 1975 meeting of the International Hydrogen Energy Society, J. W. Powell, nuclear scientist from Brookhaven National Laboratories, predicted that fusion reactors could be on line by 1995. He indicated that what was needed more than anything was funding to build a large demonstration plant. If the more than $300 million being squandered on the breeder this year and the next and the next could be released to build a demonstration plant, fusion could be here in substantial capacity before the end of the century.

During the investigations leading to this Report it was discovered that hydrogen-burning MHD, using no fossil fuels, could be a viable alternative to the conventional or breeder reactor. If the money spent on the breeder could have been spent on this technology, these plants would have been available for large scale expansion today. As it is, the hydrogen burning MHD cannot be ready before 1985.

In short, nuclear fission cannot be the near term solution to the serious energy crunch, which will begin about 1978 and begin to hurt in earnest in about ten years after that.

SECTION 4. RADIOACTIVE ISOTOPE GENERATORS

With all of the uproar about radioactive wastes it is surprising that more effort has not been made to dispose of them by using them as power sources. Such an application has recently been developed in West Germany and offered in this country by Siemens Corporation (136 Wood Ave. South, Iselin, N.J. 08830). The trade name for this radioactive isotope power supply is "Tristan."

The energy source for the Tristan unit is the radioisotope strontium 90. The heat from the radioactive decay is converted by means of thermoelectric semiconductors to electrical energy. Models with ratings from 1 to 150 watts are available. The unit, being lead shielded, has a radioactive emission of less than 200 rem/h. Underwater and land based models are available.

This power supply is capable of unattended operation for five years. The 150-watt, undersea model delivers power at 5¢ per watt hour, and the land based models for about half that amount.

SECTION 5. THE WILDEST IDEA YET

Nitinol is a nickel titanium alloy with the interesting property of having a "memory." A shape can be imprinted in the metal by heat treating it at fairly high temperature. When it comes back to ambient temperatures the shape can be relaxed and recalled by rather small temperature changes. This interesting effect was discovered by W. Buehler of the Naval Ordinance Laboratory in 1963.

This material is fairly soft in the cold phase and as strong as stainless steel in the warm phase. It can be made into an engine by using either a spring memory or a stretch memory in a set of wires or ribbons and attaching the wires or ribbons to a crankshaft. The cool phase and the heat phase are transferred to the nitinol by water. The temperature change required for initiating the phase change is 100 degrees. The range over which the phase changes occur is controllable by varying the alloy composition. It works very well at normal ambient temperature, which means that it is an excellent engine for solar energy applications.

Working models of this engine have been made by Lawrence Berkeley Laboratory. Rigway Banks and Paul Hernandez of this Laboratory demonstrated the engine at the Solar Cooling for Buildings Workshop in 1974. A 1 Kw engine would be 24 inches in diameter and 6 inches high. These investigators expect to bring the cost down to the point where the materials will be about $75 per Kw.

Whether there is a fatigue limit to nitinol is not yet known. It is thought that this alloy does not have the kind of elastic deformation that produces work hardening as in most metals. Banks and Hernandez have cycled the material to 10^7 times with no apparent change.

14

EARTH-DERIVED ENERGY RESOURCES

The three earth-derived non-fossil energy resources are hydro, geothermal and tidal. Hydro will not be discussed in this Report.

STEAM FROM THE INTERNAL FIRES

The thickness of the earth's crust down to the eternal hot magma is about 20 miles on the average. From the mountainous nature of the earth's surface it is evident that the crust has been subjected to much strain. At certain weak places the magma has pushed relatively close to the surface, and deep faults in the rocks have allowed water to seep to the hot rocks. The heated water, forced to the surface, is manifest as hot springs, geysers and fumaroles. These natural curiosities led some to expect that a heat engine could be driven by this natural steam, and steam wells have been developed to exploit this power.

These wells produce either dry steam or wet steam. Wet steam wells produce very hot water, and steam for turbo-generators is flashed off by reducing the pressure. Lately, a dry hole type well is being investigated. Locations have been identified where very hot rocks are found relatively close to the surface, and it is suspected that these rocks contain no water and produce no steam in themselves.

The oldest geothermal power installation, started in 1904 near Larderello, Italy, now has a capacity of 380 Mw from dry steam. New Zealand has installed 175 Mw capacity at the Wairaki field, and Mexico has installed a 75 Mw plant. The latter two use wet steam. The only operating geothermal field in this country is "The Geysers," which is operated by Pacific Gas and Electric Company, the largest producer of geothermal power in the world. By 1977 their installed capacity is expected to be 900 Mw.

The rights to a steam well are obtained by leasing, and it is likely that, with the rush of interest in this resource, nearly all of the promising fields are now under contract. Exploitation of this resource will probably follow a pattern similar to that used by the oil industry, where an entrepreneur drills the wells and sells the fuel to the utilities.

"The Geysers" Project

In the mid-1950s Magma Power Company of Los Angeles and Thermal Power Company of San Francisco acquired the geothermal rights to "The Geysers," an area of hot springs and fumaroles about 30 miles north of San Francisco. Six successful wells were drilled having a total output of 300,000 lb/hr at 115 psig. PG&E evaluated the output and agreed to buy steam from the drillers. In 1960 the first geothermal generator, a 12 Mw unit, went into operation. Additional units have been steadily added since that time. Unit 11, just completed, has a capacity of 106 Mw. Units 12 through 15 are expected to be in operation by 1977.

The steam drillers provide the piping from the well heads to the generating units. The supply line to the 55 Mw unit, for example, is 36 inch 3/8 inch wall carbon steel pipe. It is typically connected to 7 steam wells.

Centrifugal separators are installed in the steam pipes to remove particulate matter and moisture. The steam contains about 1% noncondensate gases consisting of carbon dioxide, ammonia, methane, hydrogen sulfide, nitrogen, argon and hydrogen. It contains a powderlike dust that deposits out on the turbines. The build-up has caused some blade failures.

The steam as it comes from the wells is not corrosive. Carbon steel can be used in this area. The turbines do not require corrosion resistant materials. At the condenser end the hydrogen sulfide oxidizes to a weak sulfuric acid. Corrosiveness increases, and austenitic stainless steels, aluminum or epoxy-fiber glass are used.

Hydrogen sulfide in the air caused serious problems in electrical contacts, relay springs and exposed wiring because it is corrosive to copper, copper alloys and silver. Tin alloy coatings were found to resist corrosion, but they were not satisfactory on current-carrying contact surfaces. Heavy galvanizing and epoxy paints have proved effective on other metal surfaces. In later installations the relays, communication equipment, 480-volt switchgear and generator excitation cubicle were placed in a clean room maintained at positive pressure with clean air from an activated carbon filter.

Close coordination between the well operators and PG&E are required to start, load and secure a generating unit. The well operators require a 24-hour notice before start-up. A shut-in well requires several hours to get it up to rated flow and to clear it of water and debris. Additional time is required to warm up the extensive steam piping and drain the condensate from it. Fast start-ups have resulted in fouling of turbines with well debris. Wells are cut sequentially as the load increases. If there is the possibility of water in the steam, the turbine is tripped to prevent damage. Whenever a unit does trip, steam is released to the atmosphere through pressure control valves. The steam jet ejectors create high noise levels, which require the use of ear protection.

Sudden changes in the flow rate and pressure drop are damaging to the wells at The Geysers. Continual load changes impose great difficulties on the steam suppliers. It is not known how true this situation may be at other geothermal locations. For these reasons, and because of the elaborate start-up procedures, the generators are operated at full load continually as base load units, not as peaking units.

The Geyser plants are run unattended except for roving operators. With the first four units roving operators were on duty only during normal working hours. Starting with units 5 and 6 an operating headquarters was established, and the roving operators work out of this station on 24-hour coverage. Annunciator alarms for each unit are transmitted to the central station. Each unit has up to 60 annunciator points. Each point has two contacts, one for trips and one for trouble points. The supervisory system has four control functions to each remote unit: one raise-lower for controlling turbo-generator output, one raise-lower for control of generator excitation, and two on-off controls for breaker contacts. The annunciators have their outputs arranged in groups consisting of trip or alarm points that are similar in nature. These groups are further classified as to routine or urgent.

The Ultimate Geothermal Capacity

No one has a reliable estimate of the maximum capacity of The Geysers and the surrounding region. Guesses run from 1000 to 2000 Mw.

No one hazards a guess as to how long a geothermal field will produce. Can it go on producing into the indefinite future, or will it eventually become depleted? It it becomes depleted, can it be rejuvenated? The fact that the Larderello fields are still producing since 1904 suggests that the life expectancy of a field is very long.

It is interesting to note that PG&E expects to get only 10% of its demand from The Geysers.

Wet Steam Resources

There appears to be a very large geothermal resource in wet steam in the Imperial Valley of California. Efforts are presently under way to develop this field. The biggest problem for wet steam technology is how to dispose of the brine after the steam has been extracted. At present the solution seems to be to re-inject it into the field.

Hot Rock Resources

During the course of a regional heat-flow study in 1969, using holes drilled for mineral exploration, David D. Blackwell discovered a unique geothermal area near Marysville, Montana. There were no surface manifestations of abnormal subsurface temperature. Blackwell speculated that the source of the high heat flow discovered is either an unexposed reservoir of thermal fluid, or a shallow, still-cooling magma chamber.

Under contract from National Science Foundation, a test hole 6000 feet deep is being drilled at the Marysville site. The working assumption is that hot rock of 700 to 800 degrees F will be discovered. Most speculation is that the rock is dry. If the rock is wet, the task will be to harness the steam. If the rock is dry and fractured, the task will be to introduce water from the surface to create steam. If the rock is dry and unfractured, various methods will be tried to make it permeable to injected water.

The outcome of this test drilling will be awaited with interest. If it is successful, it will suggest that many hidden geothermal fields remain to be discovered.

By the time this Report is available a report describing the results of this drilling will have been issued by National Science Foundation.

Geothermal Pollution

Of all of the non-fossil energy resources geothermal is most guilty of causing pollution.

The water and steam from the wells bring to the surface a number of minerals and compounds that upset the ecological balance.

At The Geysers the excess water was originally drained into Big Sulfur Creek, until it was found that the mineral content of the water was detrimental to fish life. Now the excess water is returned to the well operators for reinjection into the steam reservoir. At first there was some concern on the part of the well operators that this water might quench the wells. However, that problem did not materialize. It is now thought that reinjection may extend the productive life of the wells; there may be more heat in the reservoir than there is water to extract it.

Hydrogen sulfide is another pollution problem. With the many quadrillions of pounds of steam being brought to the surface each hour in a large field such as The Geysers many tons of hydrogen sulfide enter the atmosphere each year. Hydrogen sulfide is released in two ways: Most of it dissolves in the cooling water in the direct contact condenser, and it is then stripped out in the cooling tower. The rest is removed along with other gases by the condenser off-gas removal equipment and discharged into the air.

PG&E now carries out a hydrogen sulfide abatement program. The salts of either iron or nickel are added to the cooling tower. The salts catalyze the oxidation of hydrogen sulfide to elemental sulfur. As sulfur deposits in the cooling water system, the condenser off-gases, which contain enough methane and hydrogen in addition to the hydrogen sulfide to be combustible, are incinerated. The resulting sulfur dioxide is scrubbed out in a column using cooling water. The cooling water becomes slightly acid, requiring neutralization.

Bleeding steam from the wells prior to delivery to the turbines releases pollutants into the atmosphere, but control of this source has not yet been accomplished.

Noise pollution is recognized as one of the harmful factors of modern life unless it is abated. Geothermal fields are noisy, especially from the bleeding of the steam wells.

Geothermal resources are frequently found in regions of high scenic value. Many factions of society have come to think that deterioration of visual surroundings is too high a price to pay for technological advance.

While some of the pollution problems of geothermal power can be controlled, it will probably never be as clean as solar and wind power.

Some have raised the question whether geothermal operation might trigger seismic activity. The Department of the Interior has suggested that this hazard is serious enough to warrant careful monitoring. It is a factor about which there is too little present knowledge.

TIDAL POWER

The hassle involving the Bay of Fundy tidal power project has left a bad feeling on the subject of tidal power. The tides are supposed to offer too little gain at too much cost.

Suppose the difficulties with tidal power could be blamed on the think-big approach. Would the think-small approach make tidal power more attractive?

In the town of Cohasset, Mass., there is a tidal flat, shown in Figure 14-1, of about 14 acres. The tidal water flows through an inlet about 70 feet wide, and through this narrow neck about 6 million gallons of water flow four times a day.

Converting this flow to foot-pounds, it can be shown that this tidal flat could operate a 1200 horsepower water wheel. If the water wheel could extract 40% of the potential power of the tide, it could generate 4300 Kwhr per day, which is enough to supply 172 households. It could supply 11% of the community's household electricity. It should be capable of supplying electricity at about 8 mills per Kwhr.

An installation of this sort should be as simple as possible. No elaborate dam or other construction should be made. Other than the development of a reversible water wheel, no innovative approaches are required.

Admitting the proposition that a community ought to be self-sufficient in its residential and commercial requirements for thermal and electrical energy, then every bit of energy capacity helps. It has been shown that by intensive cultivation of space the simple flat plate collector can generate a surplus of energy beyond the community needs.

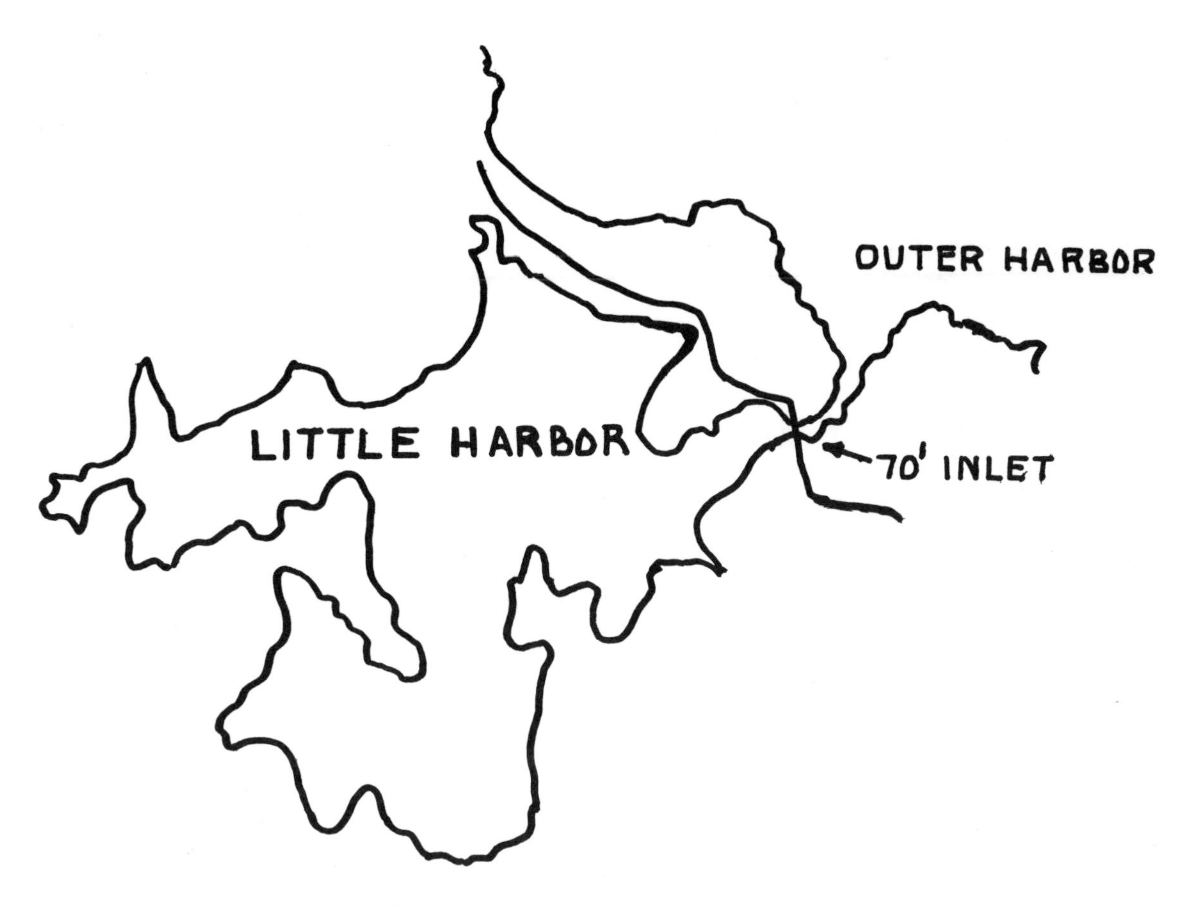

Figure 14-1. A tidal flat power basin.

However, in most communities the intensity of flat plate cultivation rapidly reaches a limit imposed by the encroachment of neighbors and such questions as how many trees should be cut down. Every bit of non-collector capacity reduces the collector density. A community that can supply 20% of its power from the wind can cut down on the number of flat plates. If it can also supply 10% of its needs from the tides, it is much more fortunate.

In a community limited for flat plate space any power source, even one supplying only 1% of capacity, is worth exploiting.

15

A LOOK AT THE FUTURE

TIME TABLES SHOW WHAT TO EXPECT

A summary of energy developments leads to the following possible time table:

Table 15-1. Non-Fossil Primary Energy Resource Time Table

Resource	Demonstration	First Production
Synthetic gas and oil from coal	1985	
Flat Plate Collectors	Done	Done
Concentrator Collectors	1978	1980
Solar Thermal Systems	1980	1983
Solar Electricity		
Cadmium Sulfide	1975	1976
Silicon	1975	1982
Large Windmills	1975	1978
Small Windmills	Done	1976
Ocean Thermal	1980	1985
Geothermal*	Done	?*
Biomass	Done	Done
Nuclear		
Breeder†		
Fusion	1990	1995

---

*GEOTHERMAL is already being exploited at The Geysers. How soon wet steam and hot rocks can be exploited is not known.

†The BREEDER is written off in this Report as a bad dream. Its development is not entered in the table.

Since solar and wind energy require storage and a secondary generation of power, progress in storage is as important as the progress in the primary resource. The status is summarized in Table 15-2.

Table 15-2. Storage and Secondary Energy Resource Time Table

Method	Demonstration	First Production
Hydrogen Electrolysis	Done	1985
Thermo-chemical Hydrogen	1985	1995
Hydrogen Fuel Cells	Done	Done
Hydrogen MHD	1985	1990
Hydrogen Storage	Done	1980
Pipe Line Hydrogen Transmission	Done	1985
Superflywheel Storage	1980	1985
High Temperature Batteries	1980	1985

The possible penetration of non-fossil fuels by the year 2020 is shown in Table 15-3.

Table 15-3. Possible Energy Sources in 2020 in Quadrillion BTU

Source	Sensible Heat	Electricity	Hydrogen
Collectors	34.6	32.2	
Solar (Energy Factory)		10	40*
Windmills		5	20†
Ocean Thermal		1	
Geothermal		0.5	
Fusion			8.8
Bio-mass	4		
Synthetics (From Coal)	26.7		

---

*Represents 12,000 square miles of solar farm.

†Windmill hydrogen will go to MHD and fuel cells.

This table optimistically assumes that fusion plants can be built faster than breeder plants. The environmental and land use problems of fusion can be significantly less than those of breeder plants. The only radioactive element is tritium, which has a half life of 10 years. If some does escape into the air, the contamination will not measureably increase the ambient radioactivity. It is hoped by 2020 that plant builders will have become rational enough to recover the low quality waste heat with Rankine cycle generators or some other device. If so, the cooling water problem as it relates to site selection will be lessened.

If 5000 Mw plants are possible by 2020, it would require 65 plants to supply 8.8 quadrillion BTU. It would require 195 plants to supply the 26.7 quadrillion BTU anticipated from coal. If the nation's needs were 180 quadrillion BTU, it would require 1318 plants of 5000 Mw size to furnish the supply. Thus, even though fusion is offered as an abundant source of energy, the task of building the capacity is formidable. For a long time to come it will make sense to split the load among a variety of other non-fossil resources.

A NEW PREDICTION OF THE ENERGY MIX

Figure 1-1 at the beginning of this Report reproduced a diagram commonly used to represent the predicted consumption of energy by fuels. This diagram was based on the assumption that the resources would primarily be fossil fuels and nuclear reactors. As a result of the investigations leading to this Report, a new diagram is suggested in Figure 15-1.

WHOSE BANDWAGON?

Two facts stand out about the energy picture. First, there is no scarcity of non-fossil energy resources to exploit. Second, most of the new resources, except windmills and collectors, will not be ready for the market until the mid-1980s.

Since there is such a large variety of resource offerings, utility managers are not inclined to commit themselves to any one until the winner emerges. However, it is suggested here that the triumph of a single fuel, as with the case of oil, will not occur again for 50 years, if ever. During the intervening 50 years it is likely that the utilities will use a much wider mix than they have ever used before.

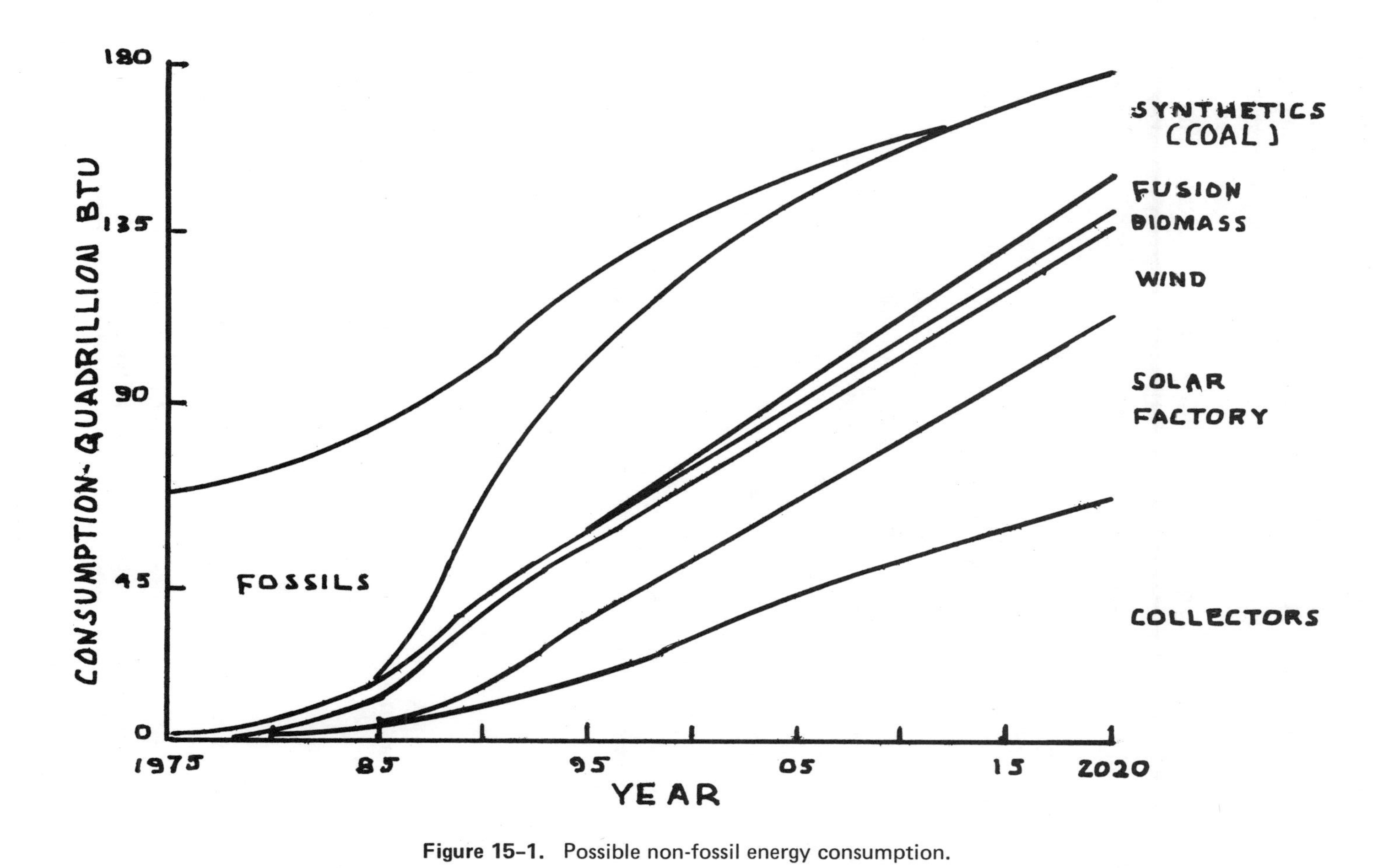

Figure 15-1. Possible non-fossil energy consumption.

If fusion fulfills its promise of unlimited energy by the early part of the 21st century, what happens to the energy mix already in operation? Will fusion supplant all other energy conversion systems? As suggested above, it will take a long time to build enough fusion plants to supply the energy demand.

The most immediate consequence of fusion may be the triumph of the energy factory concept. The industrial revolution killed the cottage industry. Will the energy factory also kill the alternate resources now being so intensively developed? There is precedence for it. The rural electrification act killed the small windmill industry in this country.

Whether the energy factory will eliminate the diffuse energy resources depends partly on how viable becomes the suggestion for local autonomy. The concept advanced in this Report is that there should be regional self-sufficiency in thermal and electrical needs for homes and commerce. The energy factory should be called on to supply power for industry, for the central business district and for transportation. Local needs should be met from solar and wind energy plus other resources that could be exploited, such as tides and pyrolysis. If these indigenous resources can be made reliable enough and cheap enough, there is no reason they should not supply most of a community's needs.

If the indigenous resources cannot be made inexpensive and reliable by the think-small approach, then the energy factory will eliminate the concept of local autonomy.

The Windpower Bandwagon

While most of the alternatives to fossil fuel will not be ready to exploit until after 1980, the utility manager's fuel pinch is now. From Table 3-1, the small windmill in about the 10 Kw size is the only item that can be planned at once.

Consider the model community. To generate 10% of the community's needs, about 250 small windmills would have to be installed, assuming sites providing rated output on the average of 10 hours a day could be found.

By waiting until 1980 it might be possible to install windmills in the megawatt size, in which case it would require only three units to provide 10% of the demand.

Not every utility can find sites suitable enough and in quantity enough to make a significant dent in the fuel pinch, but a wind survey should be taken to be sure.

The Sunpower Bandwagon

At least three utilities (Florida Power & Light Co., Tampa Electric Co. and Florida Power Corp.) are looking into the possibility of getting into the solar energy business as a sales agent or leasing agent for water heaters. While the solar water heater used alone is usually a disappointing experience to persons accustomed to the electric or gas water heater, a utility that gains experience by this method will be prepared to handle complete home heating and cooling systems later.

The electric utilities are often suggested as the logical companies to install, guarantee and service the electrical solar arrays. When the solar thermal and the solar electrical collectors are combined in the same array, either one company installs and services the collectors or there will be disputes between the electricians and plumbers and carpenters. Since the utility would be wise to be involved in solar electricity, it would also be wise for them to start now to become acquainted with the other problems of solar installation.

Since the concentrator electro-thermal panel will not be ready before 1980, and since space cooling combined with space heating and hot water will not be ready before 1980, it is recommended that the utilities use the interim to get ready to get on the sunpower bandwagon.

The Utility Planning Staff

For at least half of the time period covered by this Report energy will be in short supply unless every effort is made to exploit all possible resources. Since windpower and sunpower are abundant, easy to convert, inexhaustible and non-polluting, the priority of these resources is easy to justify.

It is therefore recommended that a utility have a planning staff that can become expert in matters pertaining to windpower and sunpower.

It is also recommended that the planning staff include a meteorologist who can take expert wind surveys and collect adequate solar insolation data.

The Government's Predictions on Solar Energy

The NSF/NASA Solar Energy Panel report in 1972 predicted solar thermal panels on 10% of the new buildings in 1985, 50% in 2000 and 85% in 2020. This Panel was not optimistic about retrofitting old buildings. Further, this growth rate does not see solar energy as a very exciting bandwagon.

Since 1972 there has been a new emphasis and a greater desperation to fire the drive for solar energy. There have also come along the exciting innovations of the concentrator collector and the combined thermal and electrical panel. The NSF Panel was unable to anticipate these innovations in 1972, and no similar predictions have since been made.

This Report suggests a more optimistic prediction: 25% of new buildings and 10% of old buildings by 1985 and 95% of new buildings and 50% of old buildings by 2000. This prediction assumes that the solar panels will supply both thermal and electrical energy.

This prediction assumes that by 1985 the home owner, and the professional and business man, will consider it the thing to do to get on the solar energy bandwagon. It also assumes that nothing happens to keep the home owner, professional and business man from deciding to use his own energy system.

16

UTILITY RESPONSE TO NON-FOSSIL ENERGY

WHAT WILL SOLAR AND WIND ENERGY DO TO THE UTILITIES?

There are mixed emotions among utility managers about solar panels. Since buildings equipped with these panels will certainly enjoy solar cooling, the utility summertime peaking problems will be reduced. Since the solar panels will eliminate electrical heating and air conditioning and water heating to a great extent, the utility's fuel oil supply problem will become more manageable. But these panels will also generate electricity. The utilities will lose revenue because the solar panels will be doing what the utilities used to do for electric heating, air conditioning, lighting and appliances.

Residences consume 30% of the distributed electricity and contribute 40% of the utility revenue. The ratio between new buildings and old buildings cannot be reliably predicted for 1985. Hence, the percent of total residences equipped with solar panels cannot be predicted. For estimating purposes, assume 20% of the residences will be so equipped. As these residences will, according to prevailing concepts, supply only 80% of their needs, the loss to the utilities will not be total. If the adoption of solar panels proceeds as predicted, by the end of the century the loss from residences alone will approach 30% of the utility revenue.

During this time commercial and public buildings will also be adopting solar panels. It is possible that by the end of the century the utilities can lose 50% of their revenue to the sun.

The Decline of the Utilities

By 1985 the concept of the energy factory will probably begin to take effect. At the same time, finding oil and gas for the conventional generators will become increasingly difficult, and more and more plants will be retired without replacement. It will become easier to buy power from the energy factory than to generate it.

At some point it will become increasingly unattractive for the utility to supply solar buildings with 20% of their power. It will not be worthwhile to extend power lines into new developments. The utility could find itself supplying more peak load than base load because the solar buildings would be supplying their own base load. The solar panel customers, en masse, could do peculiar things to the utility's load. For example, a prolonged cloudy spell could deplete the solar storage system, and the utility would be called upon for several days to supply 100% load instead of 20%. When the sun returned there would follow several days of no demand at all. The present concept of reliability, where the probability of power failure should be, say, once in 10 years, would become meaningless.

In such a case the utility would have to raise the price of supplemental power to the point where it would become profitable for the building owner to incorporate 100% capacity in his solar panel installation. Having lost by now much of its generating capacity, as noted above, the utility would be reduced to distributing power from the energy factories to the heavy industries.

It is also possible that the energy factory system will supply power in such abundance and so cheaply that there will be less incentive to adopt solar power. The energy factory system could cause the solar panel bandwagon to throw a wheel before it got started.

In either case the utility could end up being no more than a distribution system.

A SCENARIO FOR UTILITY SOLAR PANELS

The solar panel industry could threaten the economic life of the utility and upset its ability to supply reliable, reasonably-priced power. Why not, then, incorporate the solar panel business into the utility business? The question, of course, is, How?

None of the facets of the manufacture of solar panels is capital intensive compared to the oil industry, the coal industry or the automobile industry. Solar panels become very capital intensive, however, when it comes to the consumer. Numbers quoted in 1975 dollars run from $3000 to $5000 for hot water, space heating and space cooling. When solar cells are added, as they will be if the householder seriously wants to utilize solar energy, the cost would increase by another $2000. In a $50,000 house the solar energy system would probably increase the cost by 14%.

One of the significant deterrents to the spread of solar energy is that the intensive capital falls upon those least able to raise capital.

The social response to this situation is to shift the burden of raising capital to those better suited. For example, the great increase in home ownership over the last century is due in large part to the social invention of the monthly mortgage payment. The banks are expert at raising capital; the mortgage payment was the social invention that enabled the buyer to shift capital raising to the bank.

The householder could shift his solar system capital needs to a bank. Or, he could shift them to a utility, if the right social invention came along.

One such invention already exists—the machinery leasing system. The utilities already lease hot water heaters. They could extend this experience to the leasing of solar systems.

There are two drawbacks to the leasing idea, however.

One: The solar energy lease would have to include the entire system, which would consist of panels, the heat and electrical storage devices, the air conditioning and heating devices, and the controls. It would be almost impossible to terminate a lease, except through householder purchase, because to remove the system would require removing part of the roof and dismantling part of the residence interior.

Two: If the householder leased from the utility, the system would still have to obey the Lof and Tybout law of marginal costs, that is, the system should supply less than 80% of its needs. The utility would be as bad off as before; it would still be supplying only 20% of the household power.

Some utilities are toying with the idea of leasing solar systems, but if they consider these two objections they will probably give up the idea.

Another social invention is commonly employed, but, in so far as is known to this Report, the idea has not been examined relative to solar panels. This invention is the system of lease rights. It is common practice to lease the oil and mineral rights—and nowadays the hot springs rights—to a property. In the same way the utility could lease the sun rights and the wind rights on a property.

In considering the model community it was pointed out that a building has at least three solar areas, the roof, the south side wall and the grounds. It was pointed out that a community does not have to find a great open space for a solar farm—or a solar factory—in order to harvest solar energy. Buildings have these three solar areas, and any or all of them could be leased by the utility.

Assuming for a moment that the utility knew what to do with the sun and wind rights once they have leased them, what would the utility do about this 80% barrier?

First, recall what causes the 80% barrier. The solar panels and the storage system are capital intensive. The closer the system capacity approaches 100% demand the larger becomes the capital share for each additional BTU or Kwhr. Since the cost per BTU or Kwhr is made up almost entirely of capital costs, that part of a system that is used less and less becomes more and more expensive per unit output. Furthermore, no one knows for sure what 100% capacity means. What would be 100% capacity for one family would be 50% for another. What would be 100% storage for four days would be inadequate for a week of bad weather.

The utility managers, with their base load, intermediate load and peak load problems, know very well why the last 20% costs more per unit than the constant 50%. The same problem exists with sun and wind energy, except it presents itself differently. The utilities use their most efficient generators for base load and the less efficient ones for intermediate and peak loads. When they need more power they simply start up the extra generators. With the sun and wind, however, all of the energy has to be harvested and used, or stored, as fast as it develops. There is no such thing as shutting off the generator. One has to take what comes, and one has to take all that comes. If one has bought so much capacity that too much is usually produced, then the cost is too great for what is produced.

Society long ago invented a solution to this difficulty. If the harvester produces more than he needs, society helps him to sell the surplus. If he can sell his surplus, he can afford to own production capacity for more than he needs with some left over to sell to his neighbors.

If the householder had a way to sell his surplus energy, he could afford enough solar panels to supply 100% of his peak demands. If he had a way to feed his surplus backwards through his electrical meter—in effect, to use the utility grid as a storage system—he would have a way to sell his surplus. There are two insurmountables to this idea. The utility would be buying power under unacceptable conditions, and the householder would still be faced with the burden of intensive capitalization.

The only people who can sell energy are those, such as utilities, who are in the energy business. Since a utility can sell its surplus, it can break the 80% barrier.

If the utility leased the energy rights from the householder, the objective would be to sell the harvest to the householder. The energy demands of the householder would remain as variable as ever, but the utility would be able to sell his surplus to his neighbors. The utility would also be leasing energy rights from the householder's neighbors. The utility would be using the surplus of each householder to smooth out the fluctuating demands of the neighborhood.

For effective distribution, the utility would draw the energy from its lease holdings into a neighborhood Solar Station. The Solar Station would contain a storage system and a dispatching system for distributing the energy to the neighborhood customers.

The Solar Station

R. A. Fernandes (Niagara-Mohawk Power Corporation, Syracuse, New York) in a remarkable paper submitted for the January, 1975, meeting of the IEEE Power Engineering Society, shows how storage can be used by the utilities for peak shaving. He recommends a mixture of storage devices, such as super-flywheels, batteries, hydrogen electrolyzers and non-reformer fuel cells.

Each type of storage is particularly suited to some characteristic of peak shaving, or to the dynamic or static variations on the load. The response of superflywheels or batteries is as fast as switching. They may serve as spinning reserve. The superflywheel is particularly suited to quick applications of full power, or to smoothing small dynamic variations of power. It is well suited to load following applications. It is capable of fast recharging. Batteries are capable of longer discharge-recharge cycles than flywheels. Some system reserve components need to remain in the generating mode for longer periods than others. For longer service, Fernandes recommends the hydrogen electro-lyzer and fuel cell.

An important discovery that comes out of Fernandes' paper is that it takes a storage mix to accommodate all of the conditions imposed upon storage. Since the single household cannot economically provide all of the varieties of storage, dependence on a storage mix at the central Solar Station is desirable.

By combining Sandia Laboratories' idea for the solar community with Fernandes' ideas about the storage mix, a picture of what the Solar Station would look like begins to emerge.

The Solar Station becomes the neighborhood energy resource shed.

The utility will draw the energy from its collectors into the Solar Station. It is likely that each householder will maintain a heat storage tank equivalent to one day's need because such a tank will be a necessary part of his heating and cooling system. His storage costs will be kept to a minimum. Sufficient heat transfer fluid will be bypassed from the collectors to charge his one day storage.

The heat transfer fluid can be pumped from the collectors to the Solar Station with no problem. Sandia Laboratories has done a computer analysis showing that if a flow rate of 200 pounds per hour is maintained through a pipe, the loss of low quality heat in water is under 0.5 degree F for several hundred feet.

How the utility would store the heat at the Solar Station has not been assessed. A bin of eutectic salts might be used. It would also be feasible to run a Rankine cycle generator from this heat and use the power to electrolyze hydrogen. Since every conversion step involves a loss, the conversion to hydrogen and then back to heat via electricity will not be as efficient as using the heat directly. How much less efficient that route is than storing the heat and pumping it back to the household is not known.

The preferred air conditioning method for this neighborhood system is not known. Air conditioning researchers have so far confined their investigations to devices that are attached directly to the collectors. None have so far publicly announced any studies for a community type air conditioner, although the Sandia solar community investigation would be a good place to start. There is one possible advantage about air conditioning in this situation. Households that do not want it can neglect to install it without jeopardizing the economics of the Solar Station.

While it is likely that a day's supply of heat would be bypassed from the collectors to the householder's storage tank, the solar electricity should all be delivered from the collectors to the Solar Station for redistribution. Solar electricity is not satisfactory as it comes directly from the solar cells. The householder's appliances require a constant-current, constant voltage source. Solar cell current varies with solar intensity and the voltage with the load. The solar cells should charge a flywheel or a battery, or should electrolyze water into hydrogen.

This Solar Station is assumed to service a neighborhood of homogeneous residences. There would be a base load of about 50% of maximum from about 6:00 A.M. to 10:00 P.M. There would be a peak from about 7:00 to 9:00 A.M. and another and larger one from about 4:00 to 8:00 P.M., followed by a tapering to a minimum about midnight. This demand curve would prevail every day with some modifications on weekends. The demand would prevail every day regardless of the weather. If there are long stretches of poor weather, the demand curve still remains. The Solar Station's storage has to be flexible enough to handle these situations.

Analysis shows that three kinds of storage will take care of Solar Station service. They are identified by the simple terms, short term, medium term and long term. The Solar Station can respond to the neighborhood demands by supplying base load, short term, medium term and long term storage.

Base load heat would come from the daily storage in the householder's tank. Short term and medium term storage would come from the Solar Station either as sensible heat or hydrogen. Long term storage would come from the Stations's hydrogen storage.

The electrical base load would be supplied by the Station's hydrogen fuel cell. The short term peaks that occur in the morning and in the evening would be handled by the superflywheel. When cloudy days occurred, the fuel cell would continue to supply the base load. For a period up to four days the batteries would charge the flywheel each day. During those occasional extended periods when there is no sun for perhaps ten days or more, the fuel cell would charge the batteries each night during the lull period. Long term storage is called into play perhaps no more than 10% of the time during the year.

Hydrogen generation and storage plays the key role in base load supply, in long term storage, and in the proposition that local indigenous power can and should supply 100% of the neighborhood demands. An abundance of hydrogen must therefore be generated. The need for hydrogen is one argument for using the Rankine cycle generator on the heat collector to produce storage hydrogen.

If conditions permit, it would be desirable to install a neighborhood windmill to generate hydrogen.

Hydrogen can be stored in the liquid state, the gaseous state under pressure, in the solid state, or by absorption in metal hydrides. From studies and experience reported at the 1975 conference of the International Hydrogen Energy Society, storage in metal hydrides is the only practical method for an application such as the Solar Station.

The Solar Station would be two-way coupled in both electricity and piped hydrogen to the utility grid so that it would be able to sell or buy energy from the utility system.

A complete study of the Solar Station idea cannot be performed in this Report, but a preliminary estimate suggests that it could deliver reliable energy on a competitive basis.

ABOUT THE AUTHOR

Floyd Hickok is a professional technical writer and manager, now retired. Holder of an A.B. from Defiance College and an M.A. from Ohio State University, Hickok served as Technical Publications Manager for Laboratory for Electronics in Boston, Massachusetts, from the end of World War II to 1960. During the 1960s he was Technical Publications Manager for Nortronics in Needham, Massachusetts, and just before his recent retirement he was a technical writer for Raytheon Company. The author is a member of the International Hydrogen Energy Society. He was the founder and first president of the Society of Technical Writers, which later became the Society of Technical Writers and Publishers.